Trends in Logic

Volume 41

For further volumes:
http://www.springer.com/series/6645

TRENDS IN LOGIC
Studia Logica Library

VOLUME 41

SCOPE OF THE SERIES

The book series Trends in Logic covers essentially the same areas as the journal Studia Logica, that is, contemporary formal logic and its applications and relations to other disciplines. The series aims at publishing monographs and thematically coherent volumes dealing with important developments in logic and presenting significant contributions to logical research.

The series is open to contributions devoted to topics ranging from algebraic logic, model theory, proof theory, philosophical logic, non-classical logic, and logic in computer science to mathematical linguistics and formal epistemology. However, this list is not exhaustive, moreover, the range of applications, comparisons and sources of inspiration is open and evolves over time.

Roberto Ciuni · Heinrich Wansing
Caroline Willkommen
Editors

Recent Trends in Philosophical Logic

Springer

Editors
Roberto Ciuni
Heinrich Wansing
Department of Philosophy II
Ruhr-University Bochum
Bochum
Germany

Caroline Willkommen
Dresden
Germany

ISSN 1572-6126 ISSN 2212-7313 (electronic)
ISBN 978-3-319-35812-3 ISBN 978-3-319-06080-4 (eBook)
DOI 10.1007/978-3-319-06080-4
Springer Cham Heidelberg New York Dordrecht London

Printed on acid-free paper

Springer is part of Springer Science+Business Media (www.springer.com)

Preface

This volume includes some selected contributions from the conference *Trends in Logic XI*, which was held on 3–5 June 2012 at Ruhr University Bochum. The topic of the conference was *Advances in Philosophical Logic*, and indeed all the contributions (more than 80) proved to share two important features. First, they placed themselves in the methodology and topics that philosophical logic has developed in the last 20 years. Second, they pushed this methodology further: they attempted (quite successfully, in our opinion) at a solution to problems that indeed emerged from the last years of research on philosophical logic and widened the scope of application of philosophical logic to new topics and problems.

The conference proved an occasion for stimulating discussions, intellectual exchange, comparing and merging different views, reflecting on established methodology, and revealing perspectives on new directions. Also, it proved an incredible source of research outputs and results, which call for the livelihood of philosophical logic and the research in the area.

Eight out of the 12 invited papers were recently collected in Studia Logica's special issue *Advances in Philosophical Logic*. In this volume, we present a selection of the contributed papers. All of them show the high quality of current research on philosophical logic and prove a good example of the interesting directions philosophical logic may take today. Last but not least, they seem to be persuasive proof that there is progress in something philosophical. Even if we confine to the latter, that is not a bad achievement at all! We are sure that the papers collected here will give the reader a flavor of the lively and stimulating atmosphere that the contributors and participants impressed at the conference.

Finally, we thank the Ruhr University Bochum, Alexander von Humboldt Foundation, Deutsche Vereinigung für Mathematische Logik und für Grundlagenforschung der exakten Wissenschaften (DVMLG), and Gesellschaft für Analytische Philosophie (GAP). Also, we thank Judith Hecker, Andrea Kruse, Lisa Dierksmeier, and Tobias Koch for their assistance in the organization of Trends in Logic XI.

Bochum, June 2013 | Roberto Ciuni
Heinrich Wansing
Dresden | Caroline Willkommen

Contents

Semantic Defectiveness: A Dissolution of Semantic Pathology 1
Bradley Armour-Garb and James A. Woodbridge

Emptiness and Discharge in Sequent Calculus and Natural Deduction 13
Michael Arndt and Luca Tranchini

The Knowability Paradox in the Light of a Logic for Pragmatics 31
Massimiliano Carrara and Daniele Chiffi

A Dialetheic Interpretation of Classical Logic 47
Massimiliano Carrara and Enrico Martino

Strongly Semantic Information as Information About the Truth 59
Gustavo Cevolani

Priest's Motorbike and Tolerant Identity 75
Pablo Cobreros, Paul Egré, David Ripley and Robert van Rooij

How to Unify Russellian and Strawsonian Definite Descriptions 85
Marie Duží

Tableau Metatheorem for Modal Logics 103
Tomasz Jarmużek

On the Essential Flatness of Possible Worlds 127
Neil Kennedy

Collective Alternatives 139
Franz von Kutschera

da Costa Meets Belnap and Nelson 145
Hitoshi Omori and Katsuhiko Sano

Explicating the Notion of Truth Within Transparent Intensional Logic 167
Jiří Raclavský

Leibnizian Intensional Semantics for Syllogistic Reasoning 179
Robert van Rooij

Inter-Model Connectives and Substructural Logics 195
Igor Sedlár

Semantic Defectiveness: A Dissolution of Semantic Pathology

Bradley Armour-Garb and James A. Woodbridge

Abstract The Liar Paradox and its kin appear to show that there is something wrong with—something pathological about—certain firmly held principles or beliefs. It is our view that these appearances are deceiving. In this paper, we provide both a diagnosis and a treatment of apparent semantic pathology, explaining these appearances without semantic or logical compromise.

Keywords Truth · Falsity · Liar Paradox · Semantic pathology · Pretense · Deflationism · Meaninglessness · Truth-conditions · Anaphora · Truth teller · Open pair · Yablo's paradox · Understanding · Revenge problems · Pathology · T-schema · Curry's Paradox · Logical values

1 Introduction

The Liar Paradox and its kin appear to show that there is something wrong with—something *pathological* about—certain firmly held principles, or beliefs, that are either semantic, regarding the proper treatment of our aletheic predicates, or logical, regarding certain (and usually fairly entrenched) patterns of reasoning. In general, a *diagnosis* will attempt to reveal the semantic or logical principles, or beliefs, that give rise to the impending pathology, and a *treatment* will involve a modification of

B. Armour-Garb (✉)
Department of Philosophy, University at Albany—SUNY, Albany, USA
e-mail: barmour-garb@albany.edu

J. A. Woodbridge
Department of Philosophy, University of Nevada, Las Vegas, USA
e-mail: woodbri3@unlv.nevada.edu

R. Ciuni et al. (eds.), *Recent Trends in Philosophical Logic*,
Trends in Logic 41, DOI: 10.1007/978-3-319-06080-4_1,

those principles, or beliefs, which avoids the clearly unacceptable conclusions that the principles, or beliefs, tend to yield.[1]

It is our view that these appearances are deceiving. Although there are certain firmly held principles or beliefs that appear to give rise to impending semantic (or logical) pathology—a condition that threatens to manifest itself in a resultant inconsistency or indeterminacy—in fact, there is no such pathology. In this paper, we provide both a diagnosis and a treatment of apparent semantic pathology, explaining these appearances away without semantic or logical compromise.

2 Pretense and Meaninglessness

The starting point for our dissolving diagnosis of the putative cases of semantic pathology is our pretense account of "truth-talk",[2] a central aspect of which is that truth-talk (which includes the use of the falsity-predicate, as well as the truth-predicate) functions *quasi-anaphorically*, as a device of content-inheritance. But since this central aspect can be (and has been) postulated independently of a pretense account,[3] our proposed diagnosis of putative semantic pathology should be available to some other accounts of truth-talk (e.g., many deflationary accounts[4]).

The best-known instance of apparent semantic pathology is the Liar Paradox, so we will begin with this case. Our approach to the Liar Paradox is a version of the "meaningless strategy", according to which liar sentences lack content in a certain sense.[5] This status, and the reasons it applies, is what dissolves the apparent pathology of liar sentences, by blocking the threatening resultant inconsistency from ever manifesting. Given our pretense-based approach to truth, we do not hold that the content of any sentence is constituted by, or is explained in terms of, its truth-conditions. But we do maintain that there is an important sense of content that a sentence can have that involves the sentence specifying objective, worldly conditions that can obtain or not. We call such conditions, *M-conditions*. Truth-conditions are related to M-conditions, in some respects, but the former have only a thin, derivative status, as conditions for the appropriate application of the truth-predicate. On our view, the truth-conditions for a sentence are a *by-product* of its meaning, of which M-conditions are a significant component. This is in line with the *meaning-to-truth conditional* schema,

(MTC) If S means that p, then S is true iff p,

no instance of which we reject.

---

[1] See Chihara [10] on the notions of diagnosis and treatment for semantic pathology.

[2] The original version of the pretense account appears in Woodbridge [23]. For the current, improved version, see our [5, 8].

[3] See [9, 15], for explicitly anaphoric accounts.

[4] For instance, in addition to the accounts cited in Footnote 2, see the deflationary accounts given in [11, 18]. For a general account of deflationary accounts of truth, see Armour-Garb [3].

[5] See [1, 6, 12, 13] for discussion of the meaningless strategy in dealing with the Liar Paradox.

Now, while a sentence like

(1) Snow is white,

specifies M-conditions directly, others specify M-conditions only indirectly. Indeed, one of the consequences of our pretense account of truth-talk is that any M-conditions specified by an instance of truth-talk (employing either the truth-predicate or the falsity-predicate) must be a function (positive or negative) of conditions specified by the supposed content-vehicle that is putatively denoted in that instance of truth-talk.

To see this, consider a straightforward instance of truth-talk like

(2) 'Snow is white' is true.

On our view, (2) specifies indirectly just the M-conditions that (1) specifies directly. As we will show, this has an interesting consequence for liar sentences (and their putatively pathological kin): They do not specify any M-conditions.

Indeed, in the case of a liar sentence like

(L) (L) is not true,

any M-conditions that (L) specified would have to be a (negating) function of the M-conditions specified by the content-vehicle that this instance of truth-talk putatively denotes. But in this case that is "another" instance of truth-talk (in fact, it is (L) itself). This means that in order to determine the M-conditions that (L) would specify, we must look to what content-vehicle this "other" instance of truth-talk putatively denotes. This multi-step determination process can "ground out", but in the case of (L) it repeats endlessly, with the result that (L) never manages to specify any M-conditions. In a sense, we get instructions that can never be completed. Accordingly, in the "specification of M-conditions" sense of content that we intend here, no content ever manages to attach to a liar sentence like (L).

The foregoing analysis of (L) extends immediately to another familiar case of apparent semantic pathology, viz., that exhibited in the truthteller sentence

(K) (K) is true.

Here too we get an endless looping in the M-conditions determination process, with the result that (K) never manages to specify any M-conditions. In addition, because we take the falsity-predicate to involve the same sort of indirect specification of M-conditions as the truth-predicate (albeit with a negating function), the same analysis of meaninglessness also applies to what we might call a "simple liar" sentence, such as

(SL) (SL) is false.

More complicated "multi-sentence" cases get the same diagnosis. In the familiar case of a liar loop, such as

(A) (B) is false
(B) (A) is true,

each sentence is an instance of truth-talk, and so each looks to another content vehicle for any M-conditions it might specify. As it turns out, (A) and (B) each look to the other to provide M- conditions, with the result of more endless looping—albeit with a slightly wider loop—and a failure of either sentence to specify any M-conditions.[6] The same explanation applies to the related truthteller pair,

(A') (B') is true
(B') (A') is true,

as well as to the basic case of what we call "open pairs",[7]

(I) (II) is false
(II) (I) is false,

and to the strengthened open pair,

(III) (IV) is not true
(IV) (III) is not true.

While all of the cases considered thus far involve a kind of looping, it should be clear that looping is just one way in which a failure to determine M-conditions can arise. Because the truth- and falsity-predicates serve only to effect indirect specifications of M-conditions, any circumstances in which attempts to specify M-conditions indirectly do not "ground out" in some direct specification of M-conditions will generate a failure to specify M-conditions. Thus, our evaluation of liar sentences as meaningless extends beyond just other looping cases, to non-looping cases, such as the truthteller sequence,[8]

(S'_1) Sentence (S'_2) is true
(S'_2) Sentence (S'_3) is true

⋮

(S'_n) Sentence (S'_{n+1}) is true

⋮

as well as to Yablo's [25] paradox,

(S_1) For all $k > 1$, sentence (S_k) is false
(S_2) For all $k > 2$, sentence (S_k) is false

⋮

---

[6] See Grover [14] for the inspiration for this explanation. As should be clear, the loop may be made as wide as one pleases.

[7] See our [4, 6, 8, 24]. Sorensen [19–21] calls this sort of case "the no-no paradox".

[8] Kripke [16, p. 693] and Grover [14, p. 597].

(S_n) For all $k > n$, sentence (S_n) is false

⋮

In both of these examples, all of the sentences in both series fail to specify any M-conditions. This situation arises from the fact that any M-conditions specified by any sentence in either series would have to be inherited from sentences later in the series. In the truthteller sequence, each sentence looks to inherit the M-conditions of the next sentence in the series, but the series never ends, so no sentence in it ever specifies any M-conditions. In Yablo's paradox, each sentence could only specify M-conditions that are a function of M-conditions specified by all of the sentences that come after it in the series. Again, because the series has no end, no sentence in it ever specifies any M-conditions. Thus, the same explanation also applies to these non-looping cases of M-conditions determination failure.[9] As such, we endorse a version of the "meaningless strategy" for dealing with putative cases of semantic pathology in general.

3 Meaninglessness and Understanding

Any meaningless strategy for dealing with the Liar Paradox and other apparent cases of semantic pathology faces an immediate objection, which arises once we recognize that, in some sense, we *understand* the apparently problematic sentences. To simplify our discussion of this objection, we will again focus on liar sentences. Now, while we do not deny that we can understand a liar sentence like (L), it is important to note that we only understand (L) *in a sense*. We claim that there are (at least) two modes of understanding and that, while we understand (L) in one sense, we do not understand it in another. Call the sense in which we do *not* understand (L), the sense that would require knowing what M-conditions (L) specifies, 'understanding$_1$'. Call the sense in which we do understand (L) 'understanding$_2$'.

We claim that if you know the form of a sentence, the meanings of the words that are contained therein and how the sentence could be used to make a genuine assertion, then you can be said to "understand$_2$" that sentence. But if you do not know the M-conditions specified by the sentence, or whether there are any or not, then, while you may understand$_2$ the sentence, you do not *understand*$_1$ that sentence.

We contend that, although we understand$_2$ a liar sentence like (L), we do not understand$_1$ that sentence, since it fails to specify M-conditions and, thus, is meaningless in the way that we have indicated. The same explanation applies to the other cases of apparent semantic pathology that also fail to specify any M-conditions.

---

[9] We will explain how to apply the diagnosis to Curry's Paradox below.

4 Meaninglessness, Denial, and S-Defectiveness

One consequence of our view that apparently semantically pathological sentences are contentless is relevant to those who propose a speech-act solution to the Liar Paradox, according to which we can deny liar sentences non-assertorically—that is, by performing a speech act, opposite (or: dual) to affirming.[10] In general, the speech act of denial is used to express *rejection*, where to reject something is to be in a mental state, opposite (or: dual) to accepting. But since rejection is a mental state, what gets rejected is not the sentence one wishes to deny; rather, one rejects *what the sentence says*, or what it *expresses*. Now, since, on our view, liar sentences lack content, it follows that they do not have anything one can reject. So, on our view, one cannot deal with liar sentences by postulating non-assertoric denial of them.

Suppose that we are right and that we cannot either deny or affirm liar sentences and their allegedly semantically pathological kin, since there is nothing that they express and, hence, nothing to accept or reject. We still face the question of how we will characterize such sentences. And, as is familiar from attempted consistent solutions to the Liar Paradox, it is at this point that *revenge problems* generally emerge. While we believe that we can address these issues and avoid the usual problems they appear to generate, due to space considerations, we shall only sketch a way of dealing with them here.

We avoid the "first wave" of revenge problems because we take no positive or negative attitude towards the putatively pathological sentences, and we neither reason to or from them, or evaluate them, in the sense of ascribing them either a *logical value* or a *truth-value*. On our account of truth-talk, liar sentences do not admit of these sorts of evaluations. In particular, given our understanding of how truth-talk functions, it does not follow, from the fact that a sentence has no content, that the sentence is not true. Rather, it follows that it is not aletheically evaluable at all.

Keeping in mind that, on our view, liar sentences (and their kin) cannot, in the relevant sense, be understood$_1$ and, thus, cannot be evaluated in the standard ways, we then face the question of how we will (semantically) characterize them. In reply, we propose the following.[11]

As a means for characterizing putatively pathological sentences, we introduce a predicate, 'is semantically defective' (henceforth, 's-defective'), which, for present purposes, is to apply to those sentences, which, while perhaps understood$_2$, have no content. More specifically, we are inclined to claim the following, by way of clarifying 's-defective':

(i) If a sentence, *S*, is s-defective, then it has nothing, by way of content, which we can accept or reject.

And, as a result,

(ii) If *S* is s-defective, then *S* is not understood$_1$.

[10] For a speech act solution to the Liar Paradox, see [17, 22].

[11] For more on this, see our [7, 8].

Moreover,

(iii) If *S* fails to specify any M-conditions—either directly or indirectly—then S is s-defective, and it is appropriate to attribute s-defectiveness to *S*.

Finally,

(iv) If *S* is s-defective, then, since *S* will not be understood$_1$, it is not aletheically evaluable, where, if *S* is not aletheically evaluable, it cannot (correctly) be assigned or denied a truth-value.

Although there is more that we might say about *s-defectiveness*, which we are importing into our vocabulary, there are two crucial points to note. First, note that 's-defective' applies directly to sentences that do not possess content, though such sentences may be understood$_2$. (Actually, it applies to sentence tokens, though the view will not end up looking like a tokenist view, at least in any interesting sense.) Second, note that, for a given sentence, *S*, if it does not specify any M-conditions at all, then *S* is s-defective. This does not count as an *analysis* of the notion of s-defectiveness, as it leaves open the possibility that there are other ways in which a sentence may be deemed s-defective, but it will do, for what follows. Let us now apply this approach to sentences that putatively exhibit semantic pathology. Once again, we will begin by focusing on how it applies to liar sentences.

5 S-Defectiveness, Apparent Semantic Pathology, and Revenge Worries

As we saw, (L) does not specify any M-conditions, which means that, by (iii), (L) will be deemed s-defective. Given the relevant instance of the T-schema, it then follows that

(3) (L) is s-defective

will be true, and, thus, given the relevant identity,

(4) '(L) is not true' is s-defective

will also be true. However, because an s-defective claim like (L) is not truth-evaluable, from the evaluation of (L) as s-defective, it does not follow that (L) is not true (and consequently, true) because the sentence '(L) is not true' is itself s-defective. The same explanation applies to the other cases of putatively pathological sentences discussed above as well.

The pressing issue for our proposed dissolving treatment of apparent semantic pathology is whether our characterization of (L) as s-defective, and the correctness of ascribing truth to a statement of that characterization, generates other revenge problems for us. To see that it does not, consider a familiar sort of revenge problem, as found in (λ) (λ) is not true or is s-defective,

which, without contradiction, cannot be evaluated as true, false or not true. (We leave it as an exercise to the reader, how any aletheic evaluation of (λ) results in contradiction.)

Now, we would characterize (λ) as s-defective prior to any threat of inconsistency. But if we do, further paradox appears immanent. For if we maintain that (λ) is s-defective then, as we have seen, we will also accept that '(λ) is s-defective' is true. But now, given that evaluation, by disquotation, or-introduction, and enquotation, we seem to be committed also to the truth of '(λ) is not true or (λ) is s-defective', from whence inconsistency appears to be unavoidable. So, are we, then, mired in paradox, having attributed s-defectiveness to (λ)?

We are not, for paradox is avoided in the case our evaluation of (λ), in virtue of the fact that (λ) does not possess any content. This—rather than *ad hoc* stipulations geared at avoiding contradiction—is why we evaluated (λ) as s-defective in the first place. Our argument for the claim that paradox is avoided in this evaluation relies on two features: (a) that (λ) is without content; and (b) that if a standard, aletheically evaluable sentence is disjoined (or conjoined or otherwise extensionally connected) with a sentence that is without content then contentfulness cannot be preserved in the resulting complex sentence. We shall now motivate both (a) and (b).

Beginning with (a), in order for our attribution of s-defectiveness to (λ) to generate paradox, (λ) would have to have content, in the sense of specifying M-conditions. But it does not have content, and here is why. For any content that (λ) would have, both disjuncts are relevant and would have to contribute. This is so because the meaning of a disjunction is a function of the meanings of its parts. So, the meaning—and, thus, the meaningfulness—of (λ) relies, at least in part, on that of its disjuncts. If one of the disjuncts lacks content, then (λ) itself does, too. Accordingly, we will show that (λ) lacks content, by explaining why the left-hand disjunct of (λ) lacks content, where, recall, a given sentence lacks content if it fails to specify M-conditions.

As was the situation with respect to (L), any M-conditions specified by the left-hand disjunct of (λ) would have to be a product of M-conditions specified by the putative content-vehicle that the disjunct denotes—which in this case is (λ) itself. Thus, in order for the left-hand disjunct of (λ) to specify M-conditions, it is required that (λ) already has determined M-conditions. But, of course, M-conditions cannot be settled for (λ) unless, or until, M-conditions are determined for its left-hand disjunct. So, for any overall M-conditions to get specified by (λ), there would have to be an impossible sort of semantic bootstrapping, which means that the process for determining what M-conditions (λ) specifies never finishes. Since (λ) fails to specify M-conditions, it follows that the left-hand disjunct does not possess any content, and, so, neither does (λ) itself. (Notice, though, that both sentences can be understood$_2$.)

Turning to feature (b) of our response to the revenge argument, here we claim that only contentful sentences may be disjoined with other contentful sentences to yield a disjunction that is, itself, contentful and thus aletheically evaluable. We will now provide support for this claim.

Although a conjunction gets its logical value from that of its conjunctive parts and a disjunction gets its logical value from at least one of its disjunctive parts, since the content of a complex sentence is a function from the contents of its parts, a disjunction gets its content from *both* of its respective parts. What this means is that the M-conditions for a disjunctive sentence will be a function of the M-conditions for each of its disjuncts. And if one of its disjuncts specifies no M-conditions, then the disjunction itself will fail to specify any M-conditions and, because of this, will possess no content.

Since the sentence, '(λ) is not true', lacks M-conditions and, thus, has no content, disjoining it with another sentence yields a disjunctive string with no content. So, even though '(λ) is s-defective' has content and is true, disjoining this sentence with '(λ) is not true', in order to form (λ) itself, yields a sentence that has no content and, thus, is not aletheically evaluable. In the terminology that we favor, because (λ) fails to specify M-conditions, we claim, by (iii) above, that (λ) is s-defective. Thus, the revenge argument cannot bootstrap (λ) into contentfulness and thereby make it evaluable as true or false (or even as not true or not false).

Our analysis of (λ) points the way to extending the diagnosis and treatment we have given to several familiar cases of putative semantic pathology to deal with Curry's Paradox and other similarly complex cases. In a Curry sentence, such as

(C) If (C) is true, then $1 = 0$,

what we have is a complex sentence involving an extensional connective, here, the conditional.[12] As in the case of (λ), any content this complex sentence might have (in particular, any M-conditions it might specify) would have to be a product of M-conditions specified by both parts of the complex sentence—by both the antecedent and the consequent. These sub-sentential parts would both have to contribute to the content (and meaningfulness) of the whole sentence, (C). The antecedent of (C) is the sentence '(C) is true'. Because this is an instance of truth-talk, for it to specify any M-conditions, it would have to inherit them from the putative content-vehicle picked out in this instance of truth-talk. That putative content-vehicle is the sentence, (C), itself. So, in order for the antecedent to specify any M-conditions and thereby have the relevant sort of content to contribute to the content of (C) as a whole, (C) as a whole would have to already have content to pass on to its antecedent, so that the antecedent could then contribute that content to the content of (C) as a whole. In short, the antecedent and the sentence as a whole are each looking to the other to provide content. But this is just another attempt at semantic bootstrapping, so the goal cannot be fulfilled. As a result, the antecedent of (C) specifies no M-conditions and is s-defective. Since the antecedent of (C) is s-defective, the result of putting it into a complex sentence via application of an extensional operator (the conditional) results in another, complex, s-defective sentence (just as disjoining the s-defective sentence '(λ) is not true' with the non-defective sentence '(λ) is s-defective' results in an s-defective complex sentence). Thus, (C), along with any other example of a Curry sentence, also specifies no M-conditions and, so, is s-defective.

---

[12] We are assuming the conditional is the material conditional, both here and in what follows.

The same result arises for multi-sentence cases involving Curry-like conditionals, for example, what we have elsewhere called the *Curry open pair*,[13]

(C1) If (C2) is true, then $\perp$
(C2) If (C1) is true, then $\perp$,[14]

as well as the asymmetric versions of the open pair we have developed, e.g.,

(V) (VI) is not true
(VI) If (VI) is not true, then (V) is not true.[15]

In each of these kinds of cases, both members of the pair end up being s-defective. In the first pair, (C1) relies in part for any content it might have on M-conditions being specified by its antecedent. But its antecedent would have to get any content it might contribute to the content of (C1) from (C2). (C2) relies in part for any content it might have on its own antecedent specifying M-conditions, but that antecedent would have to get any content it might contribute to the content of (C2) from (C1). But we have already seen that (C1) relies ultimately for any content it might have on (C2). In the second pair, (V) looks to specify (negatively) M-conditions indirectly, by inheriting them from (VI), but (VI) ends up lacking content for the same reasons that Curry sentences do, thereby leaving (V) contentless as well. Once again, in both pairs, each sentence (and sub-sentence) is looking for M-conditions to get specified somewhere else, with the result that none ever get specified. This makes the antecedents of the complex sentences s-defective, thereby making the complex sentences as a whole s-defective, and, as a result, making any sentences looking to inherit content from them s-defective as well.

6 Closing Remarks

Our analysis of the familiar cases of putative semantic pathology diagnoses them as contentless and treats them by introducing a new way of semantically characterizing them—as s-defective. This characterization is different from assigning the relevant sentences a logical value or a truth-value, at least as those notions are standardly understood. Our approach avoids revenge-problem worries via our understanding of s-defective sentences as neither aletheically nor logically evaluable, which is to say that they cannot be assessed either for truth or falsity, or for any logical values. We shall briefly explain why this is so.

In general, we ascribe truth to a sentence when we *accept* what it "says" and we ascribe falsity to a sentence when we *reject* what it "says". As noted, acceptance and rejection are mental states and are directed at the *contents* of sentences. We can thus express our acceptance of what a sentence "says" by asserting a truth-attribution to it. And we can express our rejection of what a sentence "says" either by asserting

[13] Armour-Garb and Woodbridge [6].

[14] The symbol '$\perp$' here can be read as an expression of trivialism, i.e., "everything is true".

[15] Woodbridge and Armour-Garb [24] and Armour-Garb and Woodbridge [4, 6, 8].

the negation of the sentence or by attributing falsity to it. But, for sentences that do not possess any content, there is nothing that can be accepted or rejected. Hence, and for the other reasons that we have provided, we cannot (correctly) assertorically attribute either the truth- or the falsity-predicate to such sentences.

Within a logic, we also talk about "logical values", which, if we stick with two-valued logic, will be the values, 1 and 0. Now, there are important questions about whether a sentence's having the logical value of 1 or 0 is to be identified with its having the truth-value of true or false. But these are not questions that we can address here. What is important, for present purposes, is that, whatever we take the logical values to be, we maintain that the only sentences that can have any of those values are the aletheically evaluable ones. Since we also contend that s-defective sentences are not aletheically evaluable, we therefore conclude that none of them possesses a logical value either.

This might make it seem that we are committed to two-valued logic. But we are not. It is compatible with everything that we have said that the appropriate logic to endorse has more than two logical values. But, since s-defective sentences are not aletheically evaluable, they will not be among those sentences that will be assigned any logical value. Indeed, someone concerned, for example, with the indeterminacy presented in quantum mechanics may find a reason for assigning $\frac{1}{2}$ to certain sentences. But she will still assign $\frac{1}{2}$ only to meaningful sentences—only to sentences that are understood$_1$. (This is so, even if we assign $\frac{1}{2}$ only to sentences about which we are aletheically indifferent.) So, our current proposal does not involve taking a view on which logic to endorse.

A number of consistent solutions to the liar paradox attempt to unearth or uncover some features of a natural language that had not been adequately, or correctly, recognized. Thus, one finds, for example, that the existence of sentence tokens is taken by some theorists to indicate that there is a means for semantically characterizing liar sentences without falling victim to paradox.[16] Whatever the merits of such accounts, the important point is that those theorists claim to have *discovered* a way of providing a consistent solution to the paradoxes, given only the resources that are already available in, or for, a natural language.

We are *not* trying to do that. Rather than claiming to have found, in a language like English, an expression that can be used consistently to characterize liar sentences, we are proposing a new expression, which people should or could use, in order to describe sentences that fail to yield or possess any content. Thus, we are not claiming to have solved the Liar Paradox by discovering and calling attention to this or other under-appreciated features of a natural language like English. In fact, it is completely compatible with everything we have said here that, given certain linguistic demands (e.g., regarding expressibility), our language, or, at least, our current use of that language, results in inconsistency. Understood in this way, one of our aims in this paper was to attempt to satisfy certain expressibility demands, while dissolving the apparent threat of impending inconsistency or indeterminacy, without logical or semantic compromise.

---

[16] See Goldstein [13].

References

1. Armour-Garb, B. (2001). Deflationism and the meaningless strategy. *Analysis*, *61*(4), 280–289.
2. Armour-Garb, B. (2011). The monotonicity of 'no' and the no-proposition view. *American Philosophical Quarterly*, *49*(1), 1–14.
3. Armour-Garb, B. (2012). Deflationism (about theories of truth). *Philosophy Compass*, *7*(4), 267–277.
4. Armour-Garb, B., & Woodbridge, J. (2006). Dialetheism, semantic pathology and the open pair. *Australasian Journal of Philosophy*, *84*(3), 395–416.
5. Armour-Garb, B., & Woodbridge, J. (2010). Why deflationists schould be pretense theorists (and perhaps already are). In C. Wright & N. Pedersen (Eds.), *New Waves in Truth* (pp. 59–77). New York: Palgrave Macmillan.
6. Armour-Garb, B., & Woodbridge, J. (2012). Liars, truthtellers, and naysayers: a broader view of semantic pathology i. *Language and Communication*, *32*(4), 293–311.
7. Armour-Garb, B., & Woodbridge, J. (2013). Semantic defectiveness and the liar. *Philosophical Studies*, *164*(3), 845–863.
8. Armour-Garb, B., & Woodbridge, J. (2014). *Pretense and Pathology*. Cambridge: Cambridge University Press (forthcoming).
9. Brandom, R. (1994). *Making it Explicit*. Cambridge: Harvard University Press.
10. Chihara, C. (1979). The semantic paradoxes: a diagnostic investigation. *Philosophical Review*, *88*(4), 590–618.
11. Field, H. (1994). Deflationist views of meaning and content. *Mind*, *103*(411), 249–285.
12. Goldstein, L. (2001). Truth-bearers and the liar—a reply to Alan Weir. *Analysis*, *61*(2), 115–126.
13. Goldstein, L. (2009). A consistent way with paradox. *Philosophical Studies*, *144*(3), 377–389.
14. Grover, D. (1977). Inheritors and paradox. *Journal of Philosophy*, *74*(10), 590–604.
15. Grover, D., Camp, J., & Belnap, N. (1975). A prosentential theory of truth. *Philosophical Studies*, *27*, 73–125.
16. Kripke, S. A. (1975). Outline of a theory of truth. *Journal of Philosophy*, *72*(19), 690–716.
17. Parsons, T. (1984). Assertion, denial and the liar paradox. *Journal of Philosophical Logic*, *13*(2), 137–152.
18. Quine, W. (1986). *Philosophy of Logic*. Cambridge: Harvard University Press.
19. Sorensen, R. (2001). *Vagueness and Contradiction*. Oxford: Clarendon Press.
20. Sorensen, R. (2003). A definite no-no. In J. C. Beall (Ed.), *Liars and Heaps* (pp. 225–229). Oxford: Oxford University Press.
21. Sorensen, R. (2005). A reply to critics. *Philosophy and Phenomenological Research*, *71*(3), 712–729.
22. Tappenden, J. (1993). The liar and sorites paradoxes: towards a unified treatment. *Journal of Philosophy*, *90*(11), 551–577.
23. Woodbridge, J. (2005). Truth as a pretense. In M. Kalderon (Ed.), *Fictionalism in Metaphysics* (pp. 134–177). Oxford: Oxford University Press.
24. Woodbridge, J., & Armour-Garb, B. (2005). Semantic pathology and the open pair. *Philosophy and Phenomenological Research*, *71*(3), 695–703.
25. Yablo, S. (1993). Paradox without self-reference. *Analysis*, *53*(4), 251–252.

Emptiness and Discharge in Sequent Calculus and Natural Deduction

Michael Arndt and Luca Tranchini

Abstract We investigate the correlation between empty antecedent and succedent of the intutionistic (respectively dual-intuitionistic) sequent calculus and discharge of assumptions and the constants absurdity (resp. discharge of conclusions and triviality) in natural deduction. In order to be able to express and manipulate the sequent calculus phenomena, we add two units to sequent calculus. Depending on the sequent calculus considered, the units can serve as discharge markers or as absurdity and triviality.

Keywords Sequent calculus · Natural deduction · Dual intuitionistic logic · Pseudo-constants

1 Introduction

In this article, we are interested in the correspondence between the structural features of the sequent calculus that are often neglected in comparisons with natural deduction, namely the features of empty antecedent or empty succedent of sequents. While these features are completely symmetric in the classical sequent calculus LK, this is not the case for the intuitionisitc calculus LI. This becomes apparent when considering

---

M. Arndt and L. Tranchini—Supported by the French-German ANR-DFG project "Hypothetical Reasoning—Its Proof-Theoretic Analysis" (HYPOTHESES) (DFG Schr 275/16-2). L. Tranchini—Supported by the German Research Agency (DFG grant Tr1112/1) as part of the project "Logical consequence. Proof-theoretic and epistemological perspectives"

M. Arndt (✉) · L. Tranchini
WSI für Informatik, Eberhad-Karls-Universität, Sand 13, 72076˜ Tübingen, Germany
e-mail: arnd@informatik.uni-tuebingen.de

L. Tranchini
e-mail: luca.tranchini@gmail.com

R. Ciuni et al. (eds.), *Recent Trends in Philosophical Logic*,
Trends in Logic 41, DOI: 10.1007/978-3-319-06080-4_2,

the correspondence between intuitionistic sequents and derivations in intuitionistic natural deduction NI.

In general, the correspondence of natural deduction and sequent calculus is based on the idea that formulae in the antecedent (resp. succedent) correspond to assumptions (resp. conclusions) in natural deduction derivations. Natural deduction introduction rules, those in which the logical constant figures in the conclusion, correspond to sequent calculus right rules, in which the logical constant figures on the right hand side of the conclusive sequent. Natural deduction elimination rules, in which the logical constant figures in one of the premises, correspond to sequent calculus left rules, in which the logical constant figures on the left hand side of the conclusive sequent.

In the intuitionistic case, sequents are restricted to at most one formula in the succedent (Fig. 1). We thus have:

Sequent $\Sigma \Rightarrow A$ is derivable in LI if and only if $\Sigma \vdash A$ holds in NI.

However, this does not exhaustively express the correspondence, because there is the critical case of deriving absurdity in natural deduction.

Sequent $\Sigma \Rightarrow$ is derivable in LI if and only if $\Sigma \vdash \bot$ holds in NI.

To have a perfect correspondence, the sequent calculus is usually modified by introducing the constant $\bot$ as governed by particular axioms. Although this suffices to establish a correspondence between NI and LI we will argue that a more fine-grained correspondence can be attained.

The introduction of $\bot$ usually goes together with the definition of negation as $A \to \bot$. In this paper, we will rather take negation as primitive (not only in sequent calculus, but in natural deduction as well).

In contradistinction to the empty succedent, the empty antecedent expresses the fact that all the assumptions of the corresponding natural deduction derivation have been discharged. The empty succedent corresponds to the obtaining of a specific conclusion (i.e. $\bot$), whereas the empty antecedent corresponds to an agglomerate of discharges.

In order to obtain the complete picture, we consider also the sequent calculus and natural deduction systems for dual-intuitionistic logic LDI and NDI. Dually to what happens in LI, the antecedent of sequents of LDI are restricted to at most one formula (cf. [1, 8]). A natural deduction formulation of dual-intuitionistic logic can be obtained by considering a single-premise multiple-conclusion setting in which derivation trees branch downward, as described in Tranchini [5, 6]. The correspondence between LDI and NDI can be roughly stated as follows:

Sequent $A \Rightarrow \Gamma$ is derivable in LDI if and only if $A \vdash \Gamma$ holds in NDI.

In this dual case, it is the empty succedent that corresponds to an agglomerate of discharges, specifically, conclusion discharges, side effects of the application of some

$$\frac{}{A \Rightarrow A}\ (\text{ax}) \qquad \frac{\Sigma \Rightarrow A \qquad \Theta, A \Rightarrow \Gamma}{\Sigma, \Theta \Rightarrow \Gamma}\ (\text{cut})$$

$$\frac{\Sigma \Rightarrow \Gamma}{\Sigma, A \Rightarrow \Gamma}\ (\text{WL}) \qquad \frac{\Sigma \Rightarrow}{\Sigma, \Rightarrow A}\ (\text{WR})$$

$$\frac{\Sigma, A, A \Rightarrow \Gamma}{\Sigma, A \Rightarrow \Gamma}\ (\text{CL})$$

$$\frac{\Sigma, A, B \Rightarrow \Gamma}{\Sigma, A \wedge B \Rightarrow \Gamma}\ (\wedge\text{L}) \qquad \frac{\Sigma \Rightarrow A \qquad \Theta \Rightarrow B}{\Sigma, \Theta \Rightarrow A \wedge B}\ (\wedge\text{R})$$

$$\frac{\Sigma, A \Rightarrow \Gamma \qquad \Sigma, B \Rightarrow \Gamma}{\Sigma, A \vee B \Rightarrow \Gamma}\ (\vee\text{L}) \qquad \frac{\Sigma \Rightarrow A}{\Sigma \Rightarrow A \vee B}\ (\vee\text{R}_1) \qquad \frac{\Sigma \Rightarrow B}{\Sigma \Rightarrow A \vee B}\ (\vee\text{R}_2)$$

$$\frac{\Sigma \Rightarrow A \qquad \Theta, B \Rightarrow \Gamma}{\Sigma, \Theta, A \to B \Rightarrow \Gamma}\ (\to\text{L}) \qquad \frac{\Sigma, A \Rightarrow B}{\Sigma \Rightarrow A \to B}\ (\to\text{R})$$

$$\frac{\Sigma \Rightarrow A}{\Sigma, \neg A \Rightarrow}\ (\neg\text{L}) \qquad \frac{\Sigma, A \Rightarrow}{\Sigma \Rightarrow \neg A}\ (\neg\text{R})$$

Fig. 1 The intuitionisitc calculus LI

$$\frac{}{A \Rightarrow A}\ (\text{ax}) \qquad \frac{\Sigma \Rightarrow A, \Gamma \qquad A \Rightarrow \Xi}{\Sigma \Rightarrow \Gamma, \Xi}\ (\text{cut})$$

$$\frac{\Rightarrow \Gamma}{A \Rightarrow \Gamma}\ (\text{WL}) \qquad \frac{\Sigma \Rightarrow \Gamma}{\Sigma, \Rightarrow A, \Gamma}\ (\text{WR})$$

$$\frac{\Sigma \Rightarrow A, A, \Gamma}{\Sigma \Rightarrow A, \Gamma}\ (\text{CR})$$

$$\frac{A \Rightarrow \Gamma}{A \wedge B \Rightarrow \Gamma}\ (\wedge\text{L}_1) \qquad \frac{B \Rightarrow \Gamma}{A \wedge B \Rightarrow \Gamma}\ (\wedge\text{L}_2) \qquad \frac{\Sigma \Rightarrow A, \Gamma \qquad \Sigma \Rightarrow B, \Gamma}{\Sigma \Rightarrow A \wedge B, \Gamma}\ (\wedge\text{R})$$

$$\frac{A \Rightarrow \Gamma \qquad B \Rightarrow \Xi}{A \vee B \Rightarrow \Gamma, \Xi}\ (\vee\text{L}) \qquad \frac{\Sigma \Rightarrow A, B, \Gamma}{\Sigma \Rightarrow A \vee B, \Gamma}\ (\vee\text{R})$$

$$\frac{B \Rightarrow A, \Gamma}{A \mathrel{-\!\!\prec} B \Rightarrow \Gamma}\ (\mathrel{-\!\!\prec}\text{L}) \qquad \frac{\Sigma \Rightarrow B, \Gamma \qquad A \Rightarrow \Xi}{\Sigma \Rightarrow A \mathrel{-\!\!\prec} B, \Gamma, \Xi}\ (\mathrel{-\!\!\prec}\text{R})$$

$$\frac{\Rightarrow A, \Gamma}{\lrcorner A \Rightarrow \Gamma}\ (\lrcorner\text{L}) \qquad \frac{A \Rightarrow \Gamma}{\Rightarrow \lrcorner A, \Gamma}\ (\lrcorner\text{R})$$

Fig. 2 The dual intuitionisitc calculus LDI

rule. The empty antecedent corresponds to the obtaining of a specific assumption, namely *triviality* $\top$:

Sequent $\Rightarrow \Gamma$ is derivable in LDI if and only if $\top \vdash \Gamma$ holds in NDI.

Apart from $\top$ playing the dual role of $\bot$, in NDI and LDI implication and intuitionistic negation are replaced by co-implication '$\mathrel{-\!\!\prec}$' and dual-intuitionistic negation '$\lrcorner$' (Fig. 2).

$$\frac{\Sigma, A, B \Rightarrow \Gamma}{\Sigma, A \wedge B \Rightarrow \Gamma}(\wedge L) \qquad \frac{\Sigma \Rightarrow A, \Gamma \qquad \Theta \Rightarrow B, \Xi}{\Sigma, \Theta \Rightarrow A \wedge B, \Gamma, \Xi}(\wedge)$$

$$\frac{\Sigma, A \Rightarrow \Gamma \qquad \Theta, B \Rightarrow \Xi}{\Sigma, \Theta, A \vee B \Rightarrow \Gamma, \Xi}(\vee L) \qquad \frac{\Sigma \Rightarrow A, B, \Gamma}{\Sigma \Rightarrow A \vee B, \Gamma}(\vee R)$$

$$\frac{\Sigma \Rightarrow A, \Gamma}{\Sigma, \sim A \Rightarrow \Gamma}(\sim L) \qquad \frac{\Sigma, A \Rightarrow \Gamma}{\Sigma \Rightarrow \sim A, \Gamma}(\sim R)$$

Fig. 3 The logical rules of LK

2 Emptiness in the Sequent Calculus

The rules of the sequent calculus which actually bring forth the features of empty antecedent or empty succedent have the following characteristic: they are single premise rules and: (i) in either the antecedent or succedent of their premise some formula A occurs in some context and (ii) on that side of the sequent in their conclusion the context figures alone.

If we restrict ourselves to the (classically representative) $\{\wedge, \vee, \sim\}$-fragment of the calculus LK (Fig. 3), there are two rules that satisfy this characterization, namely (~R) and (~L):

$$\frac{\Sigma \Rightarrow A, \Gamma}{\Sigma, \sim A \Rightarrow \Gamma}(\sim L) \qquad \frac{\Sigma, A \Rightarrow \Gamma}{\Sigma \Rightarrow \sim A, \Gamma}(\sim R)$$

The result of applying the right rule is that the formula A is removed from the antecedent and—suitably amended by negation—placed in the succedent. Clearly, an application of this rule in the case of $\Sigma = \varnothing$ yields a sequent which has an empty antecedent.

Apparently, emptiness of the antecedent is simply a contextual residue. Formula A is shifted into the succedent, and the resulting emptiness of the antecedent is due to the fact that context Σ in which A occurred was already empty. In this formulation of the sequent calculus it is impossible to retain the nature of the local void that is due to the removal of A, call it an 'occurrence' of emptiness. This is the reason why emptiness of the antecedent is generally an agglomerate phenomenon in the classical calculus.

The same phenomenon can be observed also in the implication right rule as soon as the language is opportunely enriched. In the classical case, this is however not particularly significant, due to the definability of implication as $\sim A \vee B$.

Conversely, an application of (~L) results in a formula A being removed from its context in the succedent. Thus, the phenomenon of emptiness can also occur in the succedent of a sequent as described for the antecedent case.

3 The Corresponding Phenomena in Natural Deduction

In sequent calculus, at each node of a derivation all assupmtions and conclusions are written anew in the sequent labelling the node. In conrast to this, each node of natural deduction derivations is labelled with just one formula. Thus in natural deduction it is not possible to move or remove formulae from assumption or conclusion position apart by incrementally developing the derivation tree by adding nodes. Consequently, there is no notion of 'no assumptions' or of 'no conclusions' that directly corresponds to the features of emptiness in the sequent calculus. Instead, natural deduction mimics these phenomena of the sequent calculus by means of two devices: discharge of assumptions and a special propositional constant.

3.1 Assumption Discharge

In the intuitionistic case, a derivation of a sequent with empty antecedent corresponds to a *closed* derivation, that is a derivation in which the conclusion does not depend on any assumptions. One of the features of natural deduction is the fact that derivations originate in one or more assumptions. Thus, the closure of a derivation has to be obtained by successively discharging those assumptions. This is obtained by means of what Prawitz [2] calls 'improper' inference rules, that is inference rules that not only yield some new conclusion, but that additionally allow for the discharge of one or more of the assumptions from which the derivation started. This side effect of assumption discharge is displayed most prominently in the rule that governs the introduction of implicative formulae, (→I), whose applications look like:

$$\dfrac{\begin{array}{c}[A]\\ \mathcal{D}\\ B\end{array}}{A \to B}\ (\to\mathrm{I})$$

The rule expresses the fact that a derivation of B that may be based on zero or more assumptions of A can be turned into a derivation of $A \to B$ by cancelling any number of those assumptions. Prawitz uses the notation $\langle\langle \Sigma, B\rangle / \langle \Theta, A \to B\rangle\rangle$, where $\Theta = \Sigma - \{A\}$, for this deduction rule.

This rule is very similar to the sequent calculus' (→R), although the latter removes exactly one occurrence of A from the antecedent. Of course, the sequent calculus' structural rules of weakening and contraction allow the introduction of a formula A or the successive contraction of multiple occurrences of A into a single one before the application of the linear (→R).

However, while the application of (→R) in the sequent calculus makes a formula disappear from the antecedent, the application of (→I) in natural deduction (properly speaking) does not make any assumption disappear, but rather changes the status of the assumptions by marking it as discharged .

For example, compare the following derivation in NI and its corresponding LI derivation.

$$\dfrac{\dfrac{[A]^1}{B \to A}\,(\to\mathrm{I})}{A \to (B \to A)}\,(\to\mathrm{I})^1 \qquad\qquad \dfrac{\dfrac{\dfrac{A \Rightarrow A}{A, B \Rightarrow A}\,(\mathrm{WL})}{A \Rightarrow B \to A}\,(\to\mathrm{R})}{\Rightarrow A \to (B \to A)}\,(\to\mathrm{R}) \qquad (\dagger)$$

In natural deduction, negation is usually defined by means of absurdity and implication. In order to establish clear correspondences, we instead use the following rule of (¬I) as primitive:

$$\dfrac{\begin{matrix}[A]\\ \bot\end{matrix}}{\neg A}\,(\neg\mathrm{I})$$

This rule obviously matches the structure of (→I) for the special case of B being the absurdity $\bot$, i.e. in Prawitz's notation the deduction rule is simply $\langle\langle\Sigma, \bot\rangle/\langle\Theta, \neg A\rangle\rangle$, where $\Theta = \Sigma - \{A\}$.

With the same caveat as was given for (→I), this corresponds to the intuitionistic sequent calculus' (¬R).

For the correspondence to properly work, we have to assume that a sequent with an empty succedent corresponds to a natural deduction derivation of conclusion $\bot$. We now turn to this.

3.2 Absurdity

It is perhaps too strong to say that natural deduction's absurdity outright mimics the empty succedent. For, contrary to the somewhat subtle mechanic of discharge which marks a formula occurrence as 'actually being absent', absurdity is a much more forthright signal. It occurs as a propositional constant that is actually inferred at some point in the construction of a derivation. However, absurdity has been treated with justified suspicion since the rule governing it, the so-called *ex falso quodlibet* rule, breaks the harmony that is exhibited between introduction and elimination rules for the other logical signs. Indeed the ex falso looks like an elimination rule for which there is no corresponding introduction:

$$\dfrac{\bot}{C}\,(\mathrm{efq})$$

Thus, there is some reason to not regard absurdity as a proper proposition. Tennant [4] proposed to treat $\bot$ as 'a sort of punctuation mark', registering that 'the derivation has reached a dead end'. Tennant's suggestion is indeed very natural if we look at $\bot$ from the sequent calculus perspective, where the empty succedent indeed marks an extremal point in a derivation. Rules requiring the presence of some formula in the succedent can no longer be applied. Therefore, in a sense, the attention must be shifted to the formulae of the antecedent.

As mentioned previously, in sequent calculus one may end up with an empty succedent by applying negation left rule. If in the intuitionistic case this has to correspond to a derivation having the constant $\bot$ as its conclusion, it must be due to an application of the natural deduction rule that corresponds to $(\neg\text{L})$, namely $(\neg\text{E})$[1]:

$$\dfrac{\begin{array}{cc}\mathcal{D}_1 & \mathcal{D}_2 \\ \neg A & A\end{array}}{\bot}\ (\neg\text{E})$$

However, only an application of cut yields the derivation corresponding to an application of $(\neg\text{E})$.

$$\dfrac{\Sigma_1 \Rightarrow \neg A \qquad \dfrac{\Sigma_2 \Rightarrow A}{\neg A, \Sigma_2 \Rightarrow}\ (\neg\text{L})}{\Sigma_1, \Sigma_2 \Rightarrow}\ (\text{cut})$$

An application of the *ex falso quodlibet* rule following immediately $(\neg\text{E})$ would correspond to extending the sequent calculus derivation with an application of right weakening.

3.3 The Dual Perspective

We restrict ourselves to a brief exhibition of the natural deduction system for dual-intuitionistic logic NDI; for further details we recommend Schroeder-Heister [3], Wansing [9], Tranchini [5, 6]. The system NDI is a single assumption and multiple conclusions calculus in which derivations are built bottom-up.[2] Implication is replaced by *co-implication* '$\prec$' which, instead of discharging assumptions, is a operator that can discharge conclusions.

Co-implication is governed by the following elimination and introduction rules:

$$\begin{array}{c}\dfrac{A \prec B}{B}\ (\prec\text{E}) \\ [A]\end{array} \qquad\qquad \dfrac{B}{A \prec B \qquad A}\ (\prec\text{I})$$

[1] As already mentioned, negation could also be understood as a defined notion, in which case the rule is simply an instance of $(\to\text{E})$.

[2] The terminology of 'assumption' and 'conclusion' is employed for the purpose of retaining the correspondences to antecedent and succedent of sequents. The terminology of 'introduction' and 'elimination' rules follows this perspective and is thereby counter-intuitive to the direction in which derivations are constructed.

As intuitionistic negation is usually viewed as implication of $\bot$, dual-intuitionistic negation can be defined as co-implication of $\top$, where $\top$ denotes the universal trivial statement, governed by the following rule:

$$\frac{C}{\top}$$

As intuitionistic negation, we will however take dual-intuitionistic negation as primitive and governed by the following rules:

$$\begin{array}{c}\dfrac{\lrcorner A}{\top}\,(\lrcorner\mathrm{E})\\ [A]\end{array} \qquad\qquad \dfrac{\top}{\lrcorner A \qquad A}\,(\lrcorner\mathrm{I})$$

Note that in dual-intuitionistic logic the picture we described concerning the relation between emptiness in the sequent calculus and natural deduction is reversed. In LDI, empty succedents correspond to NDI derivations in which all conclusions are discharged, and empty antecedents correspond to NDI derivations in which the only assumption is the propositional constant $\top$. This means that the passage from sequent calculus to natural deduction introduces again an asymmetry in how emptiness on the left and on the right side of a sequent are rendered.

4 Units in the Sequent Calculus

We have referred to the phenomena of empty antecedent and empty succedent as structural features of the sequent calculus. We have also mentioned that assumption discharge in NI can be viewed as closely mimicking this structural feature of the emptiness of sequents, in the sense that it can be seen as a structural feature of natural deduction, at least as far as it is only a side effect of an improper inference rule. On the other hand, the absurdity constant $\bot$ provides an entirely propositional correlation to the structural feature of empty succedent.

In NDI, the situation is reversed with discharge of conclusions being a structural feature and triviality $\top$ being a propositional constant.

If we were to follow Tennant's suggestion of giving $\bot$ (resp. $\top$ in the dual case) a structural meaning (and if that were feasible without incurring other problems), both calculi would address these extremal phenomena in a purely structural manner. For the remainder of this article, we shall rather go in the opposite direction. We shall present the empty spaces as having a propositional nature already in sequent calculus.

To do this we define in each sequent calculus an infinite set of *pseudo-constants*, nullary constants labelled with formulae. We will refer to pseudo-constants as *units*.

In LK, LI and LDI, the units are uniformly defined as follows (we use $\triangledown$, $\vartriangle$ and $*$ as meta-symbols for the units and for the negation of each system):

$$\triangledown_A =_{\text{def}} A \vee *A \qquad\qquad \vartriangle_A =_{\text{def}} A \wedge *A$$

In LK, where $*$ is $\sim$, $\triangledown$ and $\vartriangle$ are respectively $\curlyvee$ and $\curlywedge$. In LI, where $*$ is $\neg$, $\triangledown$ and $\vartriangle$ are respectively $\vee$ and $\bot$. In LDI, where $*$ is $\neg$, $\triangledown$ and $\vartriangle$ are respectively $\top$ and $\wedge$. The choice of symbols will become clear in the next sections.

The main reason for labelling pseudo-constants with formulae is that it will permit to interpret them as representing the discharge of some formula A in the corresponding natural deduction derivations.[3]

While the set of formulae is supplemented by the infinitely many units, one for each formula, the units must remain distinct from proper formulae. We enforce the following restriction on the use of the units in the sequent calculi we will consider: *Units cannot be connected to other formulae through applications of left and right rules for the connectives.* Therefore, formula variables A and B that occur in the premises of rules of the calculi and are connected by a connective in the conclusion must never be instantiated by units. Note that this does not apply to the cut rule, since the cut formula does not occur in the conclusion. That is why cuts on units are permitted.

We now face the task of adding the two units to the sequent calculus. We will not add the units to the intuitionistic and the dual-intuitionistic calculi right away, but instead consider their effects on the classical calculus LK. Only afterwards will we move on to LI and LDI.

4.1 Units and Negation Rules in LK

We consider LK with conjunction and disjunction rules formulated in the multiplicative fashion.

Given the definition of $\vartriangle$, instances of $(\wedge\text{R})$ with principal formulae A and $\sim A$ read as follows:

$$\frac{\Sigma \Rightarrow A, \Gamma \qquad \Theta \Rightarrow \sim A, \Xi}{\Sigma, \Theta \Rightarrow \curlywedge_A, \Gamma, \Xi}\,(\wedge\text{R})$$

[3] In this we follow Gentzen's example in the second part of the *Untersuchungen*, where he suggests the formula $p \wedge \neg p$ (for some arbitrary propositional variable p) as propositional representation of $\bot$ in the sequent calculus as part of his translation of derivations in NI into derivations in LI. Gentzen's choice of some arbitrary p is however a merely utilitarian *ad hoc* choice. If one wished to abstract over the choice of p, one could define units as second-order formulae.

$$\vartriangle =_{\text{def}} \exists\alpha.\alpha \wedge \neg\alpha \qquad\qquad \triangledown =_{\text{def}} \forall\alpha.\alpha \vee \neg\alpha$$

However, we will actually profit from labels and thereby we chose to stay within the propositional setting.

According to the definition of the pseudo-constant $\curlywedge_A$, the consequence of this inference is actually the sequent $\Sigma, \Theta \Rightarrow A \wedge {\sim}A, \Gamma, \Xi$. Note the special case of a right premise that is an instance of the axiom:

$$\dfrac{\Sigma \Rightarrow A, \Gamma \qquad \dfrac{}{{\sim}A \Rightarrow {\sim}A}\,(\text{ax})}{\Sigma, {\sim}A \Rightarrow \curlywedge_A, \Gamma}\,(\wedge\text{R})$$

Accordingly, we can *define* negation left rule through simultaneous introduction of a unit into the succedent:

$$\dfrac{\Sigma \Rightarrow A, \Gamma}{\Sigma, {\sim}A \Rightarrow \curlywedge_A, \Gamma}\,({\sim}\text{L}_{\curlywedge})$$

This definition is simply a special case of the more general unit introduction into the succedent.

Dually, negation right rule can be defined by introducing the dual unit into the antecedent. This time the introduction of the unit is the result of a disjunction left rule.

$$\dfrac{\Sigma, A \Rightarrow \Gamma}{\Sigma, \curlyvee_A \Rightarrow {\sim}A, \Gamma}\,({\sim}\text{R}^{\curlyvee}) \quad =_{\text{def}} \quad \dfrac{\Sigma, A \Rightarrow \Gamma \qquad \dfrac{}{{\sim}A \Rightarrow {\sim}A}\,(\text{ax})}{\Sigma, \curlyvee_A \Rightarrow {\sim}A, \Gamma}\,(\vee\text{L})$$

Applications of $({\sim}\text{L}^{\curlywedge})$ and $({\sim}\text{R}^{\curlyvee})$ in which Γ and Σ are empty yield sequents in which emptiness no longer occurs. In a sense, the rules give a propositional content to the empty succedent and antecedent. (In this sense we do the opposite of what Tennant suggests.)

More precisely, call $\mathsf{LK}^{\curlyvee}_{\curlywedge}$ the result of replacing $({\sim}\text{L})$ and $({\sim}\text{R})$ with $({\sim}\text{L}^{\curlywedge})$ and $({\sim}\text{R}^{\curlyvee})$ and let $\curlywedge_\Delta =_{def} \{\curlywedge_A : A \in \Delta\}$ and $\curlyvee_\Delta =_{def} \{\curlyvee_A : A \in \Delta\}$ for any set of formulae Δ. The following holds:

Theorem 1 (Units as emptiness)

1. *If* $\Sigma \Rightarrow$ *is derivable in* LK *then* $\curlyvee_\Theta, \Sigma \Rightarrow \curlywedge_\Xi$ *for some* Θ, Ξ $(\Xi \neq \varnothing)$ *is derivable in* $\mathsf{LK}^{\curlyvee}_{\curlywedge}$.
2. *If* $\Rightarrow \Gamma$ *is derivable in* LK *then* $\curlyvee_\Theta \Rightarrow \Gamma, \curlywedge_\Xi$ *for some* Θ, Ξ $(\Theta \neq \varnothing)$ *is derivable in* $\mathsf{LK}^{\curlyvee}_{\curlywedge}$.

Proof By induction on the structure of the derivation. The critical cases are those in which negation rules are applied. Details are left to the reader. □

4.2 Units and (Co-)Implication Rules in LK

As already mentioned, in classical logic implication is defined as ${\sim}A \vee B$. In standard LK the implication right rule is derived as follows:

$$\dfrac{\Sigma, A \Rightarrow B, \Gamma}{\Sigma \Rightarrow A \to B, \Gamma}\,(\to\mathrm{R}) \quad =_{\mathrm{def}} \quad \dfrac{\dfrac{\Sigma, A \Rightarrow B, \Gamma}{\Sigma \Rightarrow {\sim}A, B, \Gamma}\,({\sim}\mathrm{R})}{\Sigma \Rightarrow {\sim}A \vee B, \Gamma}\,(\vee\mathrm{R})$$

By replacing the application of (~R) with one of $({\sim}\mathrm{R}^{\curlyvee})$, we can derive an alternative right implication rule in which the movement of A from one side to the other leaves a unit as its trace:

$$\dfrac{\Sigma, A \Rightarrow B, \Gamma}{\Sigma, \curlyvee_A \Rightarrow A \to B, \Gamma}\,(\to\mathrm{R}^{\curlyvee}) \quad =_{\mathrm{def}} \quad \dfrac{\dfrac{\Sigma, A \Rightarrow B, \Gamma \qquad \dfrac{}{{\sim}A \Rightarrow {\sim}A}\,(\mathrm{ax})}{\Sigma, \curlyvee_A \Rightarrow {\sim}A, B, \Gamma}\,(\vee\mathrm{L})}{\Sigma, \curlyvee_A \Rightarrow {\sim}A \vee B, \Gamma}\,(\vee\mathrm{R})$$

It should be observed that also in this case, the "real" shift of side taking place in (→R) is replaced by an only "apparent" shift of side in $(\to \mathrm{R}^{\curlyvee})$. The occurrence of the formula A in the premise of the rule gets locked into the unit. The occurrence of A as sub-formula of $A \to B$ in the conclusion comes from the axiom which acts as premise of (∨L).

It is also worth noticing that the definition of $(\to \mathrm{R}^{\curlyvee})$ is essentially classical, in the sense that putting $\Gamma = \varnothing$ would not make the rule intuitionistically derivable: One needs multiplicity of formulae in the succedent in order to apply the multiplicative disjunction right rule. However, this does not forbid to take an intuitionistic version of $(\to \mathrm{R}^{\curlyvee})$ as primitive. Given the correspondence of NI and LI, this would enable to view the occurrence of the unit in the rule as corresponding to the discharge of the assumption A in applications of the natural deduction (→I) rule. In the next section, we will develop this suggestion in a fully-fledged manner.

Dual considerations apply to co-implication. Given the classical definition of $A \prec B$ as ${\sim}A \wedge B$, the co-implication left rule can be derived by means of (~L) and (∧L):

$$\dfrac{\Sigma, B \Rightarrow A, \Gamma}{\Sigma, A \prec B \Rightarrow \Gamma}\,(\prec\mathrm{L}) \quad =_{\mathrm{def}} \quad \dfrac{\dfrac{\Sigma, B \Rightarrow A, \Gamma}{\Sigma, {\sim}A, B \Rightarrow \Gamma}\,({\sim}\mathrm{L})}{\Sigma, {\sim}A \wedge B \Rightarrow \Gamma}\,(\wedge\mathrm{L})$$

By replacing the application of (~L) with one of $({\sim}\mathrm{L}^{\curlywedge})$, an alternative co-implication left rule $(\prec\mathrm{L}^{\curlywedge})$ can be derived in which the (now only apparent) shift of A from the succedent to the antecedent leaves a unit as trace.

$$\dfrac{\Sigma, B \Rightarrow A, \Gamma}{\Sigma, A \prec B \Rightarrow \curlywedge_A, \Gamma}\,(\prec\mathrm{L}_{\curlywedge}) \quad =_{\mathrm{def}} \quad \dfrac{\dfrac{\Sigma, B \Rightarrow A, \Gamma \qquad \dfrac{}{{\sim}A \Rightarrow {\sim}A}\,(\mathrm{ax})}{\Sigma, {\sim}A, B \Rightarrow \curlywedge_A, \Gamma}\,(\wedge\mathrm{R})}{\Sigma, {\sim}A \wedge B \Rightarrow \curlywedge_A, \Gamma}\,(\wedge\mathrm{L})$$

Again, in a dual-intuitionistic version of the rule, the unit may be viewed as standing for the discharge of a conclusion of form A resulting by applications of (≺E) in NDI. This will be properly spelled out in the next sections.

5 Introducing Units into LI and LDI

The sequent calculi LI and LDI for intuitionistic logic and its dual are obtained by imposing an asymmetric restriction on sequent contexts. These substructural restrictions can be viewed as modifying the meaning of the empty spaces on the side of the sequent arrow to which the restriction applies.

5.1 Units in LI

In Sect. 2 the rules (~R) and (~L) have been identified as the causes for emptiness into the antecedent or succedent in the classical system. In the previous section, we showed how these rules, as well as the defined rule (→R) can be derived using the units.

We now do the same in LI. As already remarked, both (→R) and its modified version displaying the unit cannot be derived in the intuitionistic system due to the restriction on the succedent of sequents. To overcome this, we will simply take implication rules as primitive.

In LI negation is denoted by $\neg$. We also use different symbols for the units, namely, $\curlyvee$ and $\perp$. The choice of $\perp$ is motivated by the fact that applying ($\wedge$I) to the same premises of ($\neg$E) yields the formula defining the unit:

$$\dfrac{\begin{matrix}\mathcal{D}_1 \\ A\end{matrix} \qquad \begin{matrix}\mathcal{D}_2 \\ \neg A\end{matrix}}{\perp}\ (\neg\text{E}) \qquad\qquad \dfrac{\begin{matrix}\mathcal{D}_1 \\ A\end{matrix} \qquad \begin{matrix}\mathcal{D}_2 \\ \neg A\end{matrix}}{A \wedge \neg A}\ (\wedge\text{I})$$

A direct correspondence of the inference on the right is obtained through the replacement of ($\neg$L) by the following rule that both shifts a negated A into the antecedent and at the same time introduces an instance of the pseudo-absurdity $\perp_A$:

$$\dfrac{\Theta \Rightarrow A}{\Theta, \neg A \Rightarrow \perp_A}\ (\neg\text{L}_\perp)$$

In serving as a marker for the empty succedent, the unit $\perp_A$ strengthens the intuitionistic requirement of sequents having to contain no more than one succedent formula to that of sequents having to contain exactly one succedent formula or otherwise a unit.

This strengthened condition entails a complication, however. In LI, an instance of weakening of the succedent enables the transition from a sequent with empty succedent to a corresponding one that contains an arbitrary succedent formula C. With $\perp_A$ filling in a formerly empty succedent, a weakening of the succedent is no longer possible, rendering (WR) obsolete. A possible remedy of the situation would be to introduce an *ex falso quodlibet* rule:

$$\frac{\Theta \Rightarrow \bot_A}{\Theta \Rightarrow C}\ (\text{efq})$$

Following Troelstra and Schwichtenberg [7], we use an axiom instead that allows the derivation of the rule above by means of a cut on the unit:

$$\frac{}{\bot_A \Rightarrow C}\ (\bot\text{L})$$

In the previous section we suggested to view the units in ($\sim$R$^{\curlyvee}$) ($\rightarrow$ R$^{\curlyvee}$) as discharge markers. This is fully substantiated in their intuitionistic versions, given the correspondence between LI and NI:

$$\frac{\Theta, A \Rightarrow B}{\Theta, \curlyvee_A \Rightarrow A \rightarrow B}\ (\rightarrow\text{R}^{\curlyvee}) \qquad\qquad \frac{\Theta, A \Rightarrow \bot_C}{\Theta, \curlyvee_A \Rightarrow \neg A}\ (\neg\text{R}^{\curlyvee})$$

In both rules, the unit $\curlyvee_A$ is retained in place of an antecedent (assumptive) formula that is shifted (discharged) in the usual intuitionistic rule. In the case of negation introduction, a further modification of the premise guarantees that its succedent contains the unit $\bot_C$ for some formula C.

Let $\text{LI}^{\curlyvee}_{\bot}$ be the calculus that is obtained from LI through the replacement of ($\rightarrow$R), ($\neg$R) and ($\neg$L) by, respectively, ($\rightarrow$ R$^{\curlyvee}$), ($\neg$R$^{\curlyvee}$) and ($\neg L_{\bot}$) as well as the replacement of (WR) by the axiom ($\bot$L).

Note that instances of the unit that serves as cancellation marker can simply accumulate in the antecedent of sequents over the course of a derivation. For example, compare the following $\text{LI}^{\curlyvee}_{\bot}$ derivation to the LI derivation (†):

$$\cfrac{\cfrac{\cfrac{A \Rightarrow A}{A, B \Rightarrow A}\ (\text{WL})}{A, \curlyvee_B \Rightarrow B \rightarrow A}\ (\rightarrow\text{R}^{\curlyvee})}{\curlyvee_A, \curlyvee_B \Rightarrow A \rightarrow (B \rightarrow A)}\ (\rightarrow\text{R}^{\curlyvee})$$

The formula $A \rightarrow (B \rightarrow A)$ is no longer derivable in the usual sense (i.e. as the succedent of a sequent with an empty antecedent), since the antecedent contains discharge markers $\curlyvee_A$ and $\curlyvee_B$.

An interesting correspondence is obtained, however, when we modify the derivability relation of NI in such a way that discharged formulae are not simply dropped from the multiset of assumptions. Instead, a single instance of the discharged assumptions (regardless of whether it is a vacuous or a multiple discharge) is collected into a second multiset that keeps track of discharged assumptions. As the derivability relation for NI is merely a notational tool, this does not change the calculus. Thus, $[\Theta]\Sigma \vdash^* A$ expresses that A is derivable from Σ in NI with discharged assumptions collected in Θ in the manner just described. A closed derivation of A in NI is then a derivation $[\Theta] \vdash^* A$ for some Θ. Hence the fact that closed derivations are seen as relative to a set of discharged assumptions makes '$\vdash^*$' more fine grained than '$\vdash$'.

Theorem 2 *Let the formulae in Σ, Θ and formula A be formulae of* NI, *i.e. unit-free, and let* $\curlyvee_\Theta = \{\curlyvee_A : A \in \Theta\}$. *Then the following holds:*

1. $[\Theta]\Sigma \vdash^* A$ *holds in* NI *if and only if* $\curlyvee_\Theta, \Sigma \Rightarrow A$ *is derivable in* $\mathsf{LI}_\bot^\curlyvee$.
2. $[\Theta]\Sigma \vdash^* \bot$ *holds in* NI *if and only if* $\curlyvee_\Theta, \Sigma \Rightarrow \bot_A$ *is derivable in* $\mathsf{LI}_\bot^\curlyvee$.

Rather than establishing this result directly, we first relate derivations in $\mathsf{LI}_\bot^\curlyvee$ to derivations in LI and thus, via the well-known correspondence, to derivations in NI. We will describe the correspondence stated by the theorem in some more detail further below.

To establish such a relation we have to get rid of residual discharge markers occurring in sequents. This can be done by adding the following axiom to the rules and the axioms that comprise $\mathsf{LI}_\bot^\curlyvee$:

$$\frac{}{\Rightarrow \curlyvee_A}\ (\curlyvee\mathrm{R})$$

Through applications of cut, this axiom allows for the removal of all units $\curlyvee_A$ that were introduced by means of $(\to\mathrm{R}^\curlyvee)$ or $(\neg\mathrm{R}^\curlyvee)$. This allows us to state the following correspondence between the intuitionistic sequent calculi with and without units.

Lemma 1 *Let the formulae in Σ and formula A be formulae of* LI, *i.e. unit-free. Then the following holds:*

1. $\Sigma \Rightarrow A$ *is derivable in* LI *if and only if* $\Sigma \Rightarrow A$ *is derivable in* $\mathsf{LI}_\bot^\curlyvee + (\curlyvee\mathrm{R})$.
2. $\Sigma \Rightarrow$ *is derivable in* LI *if and only if* $\Sigma \Rightarrow \bot_A$ *is derivable in* $\mathsf{LI}_\bot^\curlyvee + (\curlyvee\mathrm{R})$.

Proof The direction from left to right is rather straightforward. In order to translate a LI derivation into a $\mathsf{LI}_\bot^\curlyvee$ derivation, translate and compose the subderivations by following this procedure:

1. Replace every application of a critical rule $(\neg\mathrm{R})$ or $(\to\mathrm{R})$ by, respectively, $(\neg\mathrm{R}^\curlyvee)$ and $(\to\mathrm{R}^\curlyvee)$, followed by a cut with the axiom $(\curlyvee\mathrm{R})$.
2. Replace every application of $(\neg\mathrm{L})$ by an application of $(\neg\mathrm{L}_\bot)$, and, while tracing the derivation downwards from that point on, replacing empty succedents by the corresponding unit. Replace any application of (RW) by a cut with the corresponding instance of $(\bot\mathrm{L})$.
3. If any application of a rule in the LI derivation has two premises with empty succedent, and their translations into $\mathsf{LI}_\bot^\curlyvee$ have succedents $\bot_A$ and $\bot_B$, let the conclusion have the succedent $\bot_{A\vee B}$. This applies to the rule $(\vee\mathrm{L})$ the only one that can merge empty succedents.

The reverse direction is somewhat more involved, because formulae C introduced via the axiom $(\bot\mathrm{L})$ can be side formulae of further logical rules that are applied before $\bot_A$ is eventually cut. For this reason, all cuts on units $\bot_A$ have to be pushed upwards in the derivation by following the usual cut elimination procedure until a cut that has an instance of $(\bot\mathrm{L})$ as its premise is obtained. Only then can subderivations of the resulting $\mathsf{LI}_\bot^\curlyvee$ derivation be translated and composed. In doing this, the following points have to be observed:

1. Whenever one of the new rules ($\neg R^{\curlyvee}$), ($\rightarrow R^{\curlyvee}$) or ($\neg L_{\bot}$) is applied in the $\mathsf{LI}^{\curlyvee}_{\bot}$ derivation, simply employ the original LI rules.
2. Any application of cut that has one of the new axioms as one of its premises is simply dropped. □

Corollary 1 *Let the formulae in Σ and formula A be formulae of* NI, *i.e. unit-free. Then the following holds:*

1. $\Sigma \vdash A$ holds in NI *if and only if $\Sigma \Rightarrow A$ is derivable in* $\mathsf{LI}^{\curlyvee}_{\bot} + (\curlyvee R)$.
2. $\Sigma \vdash \bot$ holds in NI *if and only if $\Sigma \Rightarrow \bot_A$ is derivable in* $\mathsf{LI}^{\curlyvee}_{\bot} + (\curlyvee R)$.

Proof This follows from the preceding lemma by means of the well-known correspondence result between derivability in LI and NI. □

Proof of Theorem 2 The correspondence between the modified derivability in NI and derivability in $\mathsf{LI}^{\curlyvee}_{\bot}$ without axiom ($\curlyvee R$) is established analogously to Corollary 1.[4] The difference is that in the direction from NI to $\mathsf{LI}^{\curlyvee}_{\bot}$, applications of ($\rightarrow R^{\curlyvee}$) and ($\neg R^{\curlyvee}$) are *not* followed by the cuts removing the units. □

5.2 *Units in* LDI

In LDI negation is denoted by $\neg$. We also use different symbols for the units, namely, $\curlywedge$ and $\top$. The choice of $\top$ is motivated by the fact that applying ($\vee$E) to the same conclusions of ($\neg$I) yields the formula defining the unit:

$$\dfrac{\top}{A \qquad \neg A}\,(\neg I) \qquad\qquad \dfrac{A \vee \neg A}{A \qquad \neg A}\,(\vee E)$$
$$\mathcal{D}_1 \qquad \mathcal{D}_2 \qquad\qquad\qquad \mathcal{D}_1 \qquad \mathcal{D}_2$$

As in LI $\curlyvee$ is properly interpreted as an assumption discharge marker, in LDI $\curlywedge$ is a proper conclusion discharge marker.

Let $\mathsf{LDI}^{\top}_{\curlywedge}$ be the calculus that is obtained from LDI through the replacement of ($\prec$L), ($\neg$L) and ($\neg$R) by, respectively, ($\prec L_{\curlywedge}$), ($\neg L_{\curlywedge}$) and ($\neg R^{\top}$) as well as the replacement of (WL) by the axiom ($\top$R). Furthermore, the additional axiom ($\curlywedge$L) can be used to remove markers for discharged conclusions:

$$\dfrac{}{\curlywedge_A \Rightarrow}\,(\curlywedge L)$$

We can state the dual correspondence results.

---

[4] If one wished a direct correspondence between discharged formulae in the $*$-red derivability and formulae (and not just labelled units) in sequent calculus, one could introduce a sequent calculus LI^* which instead of introducing units labelled with the discharged formulae, just put the latter ones in a special "discharge" context.

Fig. 4 The roles of the four units in the two calculi

	∇_A	Δ_A
$\mathsf{LI}^{\curlyvee}_{\bot}$	$\dfrac{}{\Rightarrow \curlyvee_A}$ ($\curlyvee$R) discharge of assumptions	$\dfrac{}{\bot_A \Rightarrow C}$ ($\bot$L) absurdity
$\mathsf{LDI}^{\top}_{\curlywedge}$	$\dfrac{}{C \Rightarrow \top_A}$ ($\top$R) triviality	$\dfrac{}{\curlywedge_A \Rightarrow}$ ($\curlywedge$L) discharge of conclusions

Theorem 3 *Let the formulae in* Γ, Ξ *and formula* A *be formulae of* NDI, *i.e. unit-free, and let* $\curlywedge_{\Xi} = \{\curlywedge_A : A \in \Xi\}$. *Then the following holds:*

1. $A \vdash^* \Gamma[\Xi]$ *holds in* NDI *if and only if* $A \Rightarrow \Gamma, \curlywedge_{\Xi}$ *is derivable in* $\mathsf{LDI}^{\top}_{\curlywedge}$.
2. $\top \vdash^* \Gamma[\Xi]$ *holds in* NDI *if and only if* $\top_A \Rightarrow \Gamma, \curlywedge_{\Xi}$ *is derivable in* $\mathsf{LDI}^{\top}_{\curlywedge}$.

Lemma 2 *Let the formulae in* Γ *and formula* A *be formulae of* LDI, *i.e. unit-free. Then the following holds:*

1. $A \Rightarrow \Gamma$ *is derivable in* LDI *if and only if* $A \Rightarrow \Gamma$ *is derivable in* $\mathsf{LDI}^{\top}_{\curlywedge}$ + ($\curlywedge$L).
2. $\Rightarrow \Gamma$ *is derivable in* LDI *if and only if* $\top_A \Rightarrow \Gamma$ *is derivable in* $\mathsf{LDI}^{\top}_{\curlywedge}$ + ($\curlywedge$L).

Corollary 2 *Let the formulae in* Γ *and formula* A *be formulae of* NDI, *i.e. unit-free. Then the following holds:*

1. $A \vdash \Gamma$ *holds in* NDI *if and only if* $A \Rightarrow \Gamma$ *is derivable in* $\mathsf{LDI}^{\top}_{\curlywedge}$ + ($\curlywedge$L).
2. $\top \vdash \Gamma$ *holds in* NDI *if and only if* $\top_A \Rightarrow \Gamma$ *is derivable in* $\mathsf{LDI}^{\top}_{\curlywedge}$ + ($\curlywedge$L).

6 Summary

Adding the two units to the intuitionistic and dual-intuitionistic sequent calculi has shed some light on the status of emptiness in antecedent and succedent of those calculi through the axioms that are required to guarantee interderivability of sequents between LI and $\mathsf{LI}^{\curlyvee}_{\bot}$ and derivability statements for NI on one hand and between LDI and $\mathsf{LDI}^{\top}_{\curlywedge}$ and derivability statements for NDI on the other hand. Units as discharge markers are addressed by axioms which mention the units alone and may thus be viewed as purely structural. The axioms for units that represent constants relate the unit to some arbitrary formula and thus represent the *ex falso quodlibet* and its dual in the sequent calculus.

We employed four units to emphasize the difference in behaviour that depends on the impact of the structural restrictions imposed on the intuitionistic and dual-intuitionistic sequent calculi. This is summarized in Fig. 4. Note, however, that, strictly speaking, ($\curlyvee$R) is not an axiom of $\mathsf{LI}^{\curlyvee}_{\bot}$, nor is ($\curlywedge$L) an axiom of $\mathsf{LDI}^{\top}_{\curlywedge}$.

References

1. Czemark, J. (1977). A remark on Gentzen's calculus of sequents. *Notre Dame Journal of Formal Logic*, *18*(3), 471–4.
2. Prawitz, D. (1965). *Natural deduction. A proof-theoretical study*. Stockholm: Almqvist & Wiksell.
3. Schroeder-Heister, P. (2009). Schluß und Umkehrschluß: ein Beitrag zur Definitionstheorie. In: Gethmann CF (ed) Akten des XXI Deutschen Kongresses für Philosophie (Essen, 15.-19.9.2008), Deutsches Jahrbuch Philosophie, Band 3, Felix Meiner Verlag, Hamburg.
4. Tennant, N. (1999). Negation, absurdity and contrariety. In D. M. Gabbay & H. Wansing (Eds.), *What is negation?*. The Netherlands: Kluwer Academic Publishers.
5. Tranchini, L. (2010). Refutation: A proof-theoretic account. In C. Marletti (Ed.), *First Pisa colloquium in logic*. Pisa: Language and Epistemology, ETS.
6. Tranchini, L. (2012). Natural deduction for dual-intuitionistic logic. *Studia Logica*, *100*(3), 631–48.
7. Troelstra, A. S., & Schwichtemberg, H. (1996). *Basic proof theory*. Cambridge: Cambridge University Press.
8. Urbas, I. (1996). Dual-intuitionistic logic. *Notre Dame Journal of Formal Logic*, *37*(3), 440–451.
9. Wansing, H. (2010). Proofs, disproofs, and their duals. In C. Areces & R. Goldblatt (Eds.), *Advances in Modal Logic 7, Papers from the 7th Conference on "Advances in Modal Logic"* (pp. 483–505). Nancy, France: College Publications (9–12 September 2008).

The Knowability Paradox in the Light of a Logic for Pragmatics

Massimiliano Carrara and Daniele Chiffi

Abstract The Knowability Paradox is a logical argument showing that if all truths are knowable in principle, then all truths are, in fact, known. Many strategies have been suggested in order to avoid the paradoxical conclusion. A family of solutions—called *logical revision*—has been proposed to solve the paradox, revising the logic underneath, with an intuitionistic revision included. In this paper, we focus on so-called *revisionary solutions* to the paradox—solutions that put the blame on the underlying logic. Specifically, we analyse a possibile translation of the paradox into a modified intuitionistic fragment of a logic for pragmatics (KILP) inspired by Dalla Pozza and Garola [4]. Our aim is to understand if KILP is a candidate for the logical revision of the paradox and to compare it with the *standard* intuitionistic solution to the paradox.

Keywords Knowability · Logic for pragmatics · Assertion

1 Introduction

Church-Fitch's Knowability Paradox shows that from the assumptions that all truths are knowable and that there is at least an unknown truth (i.e. that we are non-omniscient) follows the undesirable conclusion that all truths are known. The paradox of knowability is considered a problem especially for antirealists on truth.

M. Carrara (✉)
FISPPA Department, P.zza Capitaniato 3, 35139 Padova, Italy
e-mail: massimiliano.carrara@unipd.it

D. Chiffi
Unit of Biostatistics, Epidemiology and Public Health, Department of Cardiac Thoracic and Vascular Science, via Loredan 18, 35131 Padova, Italy
e-mail: daniele.chiffi@unipd.it

R. Ciuni et al. (eds.), *Recent Trends in Philosophical Logic*,
Trends in Logic 41, DOI: 10.1007/978-3-319-06080-4_3,

An antirealist way of answering the criticisms consists in revising logic, assuming (for example) the intuitionistic logic as the right logic, thus blocking the paradox through the adoption of a revision of the logical framework in which the derivation is made.

We take for granted that a revision of the logical framework can be considered as the right solution to the paradox. Aim of the paper is to analyse if the paradox is reproducible within a *logic for pragmatics* (LP), specifically into a modified intuitionistic fragment of a logic for pragmatics (KILP) inspired by Dalla Pozza and Garola [4]. The basic idea of the paper is that if some epistemic aspects associated with the notion of *assertion*, which are merely implicit in some philosophical conceptions of intuitionistic logic (on this aspect see Sundholm [23]), can be explicated in a proper way in the pragmatic language, then KILP seems to be—at least *prima facie*—as good as other logical frameworks for the solution of the knowability paradox.

The paper is divided into eight sections. Section 2 is devoted to briefly outlining the structure of the knowability paradox. In Sect. 3, we sketch the intuionistic solution to the paradox. Then, an analysis of the difficulties of the intuionistic solution, specifically the *Undecidedness paradox of Knowability*, is sketched in Sect. 4. In Sect. 5, LP and ILP are introduced. Section 6 deals with an analysis of the paradox in KILP. Section 7 is devoted to a comparison between our solution and the intuitionistic one. Some provisional conclusions of the paper are outlined in Sect. 8.

2 Knowability Paradox

The *Knowability Paradox* is an argument showing that, if every truth is knowable, then every truth is also actually known. Such a paradox is based on two principles: the *principle of knowability* and the *principle of non-omniscience*. The principle of knowability KP can be expressed in the following way:

$$(\text{KP})\forall p(p \to \Diamond Kp)$$

while non-omniscience (Non-Om) is formulated as:

$$(\text{Non-Om})\exists p(p \wedge \neg Kp)$$

The expression $'Kp'$ reads "p is, has been or will be known by somebody". Assume the following two properties of knowledge:

1. the distributive property over conjunction (Dist), i.e. if a conjunction is known, then its conjuncts are also known, and
2. the factivity of knowledge (Fact), i.e. if a proposition is known, then it is true.

Assume the following two unremarkable modal claims, which can be formulated using the usual modal operators $\Diamond$ ("it is possible that") and $\Box$ ("it is necessary that"). The first is the *Rule of Necessitation*:

$$\text{(Nec) If } p \text{ is a theorem then } \Box p$$

The second rule establishes the interdefinability of the modal concepts of necessity and possibility:

$$\text{(ER) } \Box\neg p \text{ is logically equivalent to } \neg\Diamond p$$

From KP and Non-Om a contradiction follows. Fitch [7] and Church (we follow here [21]) proved that

$$(*)\, \forall p \neg\Diamond\, K\,(p \wedge \neg Kp)$$

is a theorem. But if (*) and Non-Om hold, then KP has to be rejected, since the substitution of $p \wedge \neg Kp$ for p in KP leads to a contradiction.

On the other hand, if KP is accepted, then Non-Om must be denied. However, the negation of Non-Om is equivalent to the formula asserting that "$\forall p(p \rightarrow Kp)$".

Therefore, from KP it follows that every sentence is known and this fact seems to be particularly problematic for the holders of antirealism who accept KP. This argumentation shows that in the presence of (relatively unproblematic) principles Dist and Fact, the thesis that all truths are knowable KP entails that all truths are known. Since the latter thesis is clearly unacceptable, the former must be rejected. We must conclude conceding that some truths are unknowable.

The proof of the theorem is based on the two following arguments that hold in any minimal modal system.

First argument:

(1) $p \wedge \neg Kp$	Instance of Non − Om
(2) $(p \wedge \neg Kp) \rightarrow \Diamond K\,(p \wedge \neg Kp)$	Substitution of "$p \wedge \neg Kp$" for p in KP
(3) $\Diamond K\,(p \wedge \neg Kp)$	From (1) and (2) and *modus ponens*

Second independent argument:

(4)	$K(p \wedge \neg Kp)$	Assumption
(5)	$(K\,p \wedge K\neg Kp)$	Distributivity of K
(6)	$(K\,p \wedge \neg Kp)$	Factivity of K
(7)	$\bot$	Contradiction
(8)	$\neg\,(K\,(p \wedge \neg Kp))$	*Reductio, discarging* (4)
(9)	$\Box\neg\,(K\,(p \wedge \neg Kp))$	(Nec)
(10)	$\neg\Diamond\,(K\,(p \wedge \neg Kp))$	(ER)

From (3) and (10) a contradiction follows. The result of the paradox can be summarized in the following theorem:

$$\text{(T1) } \exists q(q \wedge \neg Kq) \rightarrow \neg\forall q(q \rightarrow \Diamond Kq)$$

Furthermore, notice also that the converse of (T1) can be easily demonstrated; in fact, by the principle that what is actual is possible, we obtain the theorem:

$$(\text{T2})\ \forall q(q \rightarrow Kq) \rightarrow \forall q(q \rightarrow \Diamond Kq).$$

which is provably equivalent to the theorem:

$$(\text{T3})\ \neg\forall q(q \rightarrow \Diamond Kq) \rightarrow \exists q(q \wedge \neg Kq)$$

(T1) and (T3) validate the following theorem:

$$(\text{T})\ \exists q(q \wedge \neg Kq) \leftrightarrow \neg\forall q(q \rightarrow \Diamond Kq)).$$

If T is a theorem, by applying the *Rule of Necessitation* to T, we obtain:

$$(\text{TN})\ \Box(\exists q(q \wedge \neg Kq) \leftrightarrow \neg\forall q(q \rightarrow \Diamond Kq)).$$

Now, notice that Non-Om $\exists p(p \wedge \neg Kp)$—the non-omniscience thesis—is the result of a commonsensical observation according to which, *de facto*, actually there are true propositions that we do not know. It is not a logical principle of the paradox, nor it is introduced through a logical argument.

3 The Revision of the Logical Framework: On the Intuitionistic Solution to the Knowability Paradox

Different ways to block the knowability paradox have been proposed. They are usually grouped into three main categories:

- Restriction of the possible instances of KP.
- Reformulation of the formalization of the knowability principle.
- Revision of the logical framework in which the derivation is made.

As mentioned, we only concentrate on the last set of proposals, specifically on the intuitionistic proposal of revising the logical framework. Intuitionistic logic is considered as the right logic in an antirealistic conception of truth, a conception embracing an *epistemic* point of view on truth. A version of this epistemic conception, compatible with intuitionism, is the following one:

(A) A is true if and only if it is possible to exibit a direct justification for A.

If a justification is something connected to our linguistic capacities, namely not transcending our epistemic capacities, an antirealist can infer that:

(B) If it is possible to exhibit a direct justification for A, then it is possible to know that A.

Putting (A) and (B) together we get the knowability principle:

(KP) If A is true, then it is possible to know that A.

But, as said, from KP, every sentence turns out to be known. Supporters of an intuitionistic solution to the knowability paradox argue that

(KP) If$A$is true, then it is possible to know that A

can be weakened and formulated as a valid intuitionistic formula:

$$\text{(KPI)}\, \forall p\, (p \rightarrow \neg\neg Kp)$$

obtaining in this way a formula blocking the paradox [24]. Indeed, consider the conclusion of the paradox, i.e.:

$$\neg\exists p(p \wedge \neg Kp).$$

From the conclusion we may intuitionistically derive

$$\forall p \neg(p \wedge \neg Kp).$$

But if the double negation is not eliminated, then an instance of the above formula:

$$\neg(p \wedge \neg Kp)$$

does not entail

$$(p \rightarrow Kp).$$

It only entails KPI. An anti-realist is ready to accept KPI, provided that the logical constants are understood in accordance with intuitionistic rather than classical logic. Following Dummett [6], an anti-realist will prefer KPI to KP as a formalization of his view concerning the relation of truth to knowledge.

4 Difficulties in the Intuitionistic Solution to the Knowability Paradox

There are two connected difficulties regarding the intuitionistic revision of the logic for the treatment of the *knowability paradox.*

Firstly, according to Dummett [6], the consequent of KPI means, from an intuitionistic point of view, that "there is an obstacle in principle to our being able to deny that p will ever be known", or, in other words "the possibility that p will come to be known always remains open". From an anti-realistic point of view, the last claim holds good for every propositions p. In Dummett's opinion this is what (KPI) expresses. Observe that anti-realists (or justificationists) do not deny that there are true proposition that *in fact* will never be known,

> ... But that there are true propositions that are *intrinsically* unknowable: for instance one stating the exact mass in grams, given by a real number, of the spanner I am holding in my hand ([6], p. 52).

Now, although intrinsically unknowable propositions are difficult to be thought, one may consider the following sentence due to Pap [13] as a possible objection to Dummett's thesis (a similar sentence can be found in Poincaré's works[1]):

> Every body in the universe, including our measuring rods, is constantly expanding, the rate of expansion being exactly the same for all bodies" (p. 37).[2]

Pap's sentence is not verifiable, even if it has a definite truth-condition; namely we know how the world should be in order to make the sentence true. This point was also envisaged by Russell (in [22]).[3] If we accept such analysis of Pap's sentence we obtain a case where it does not happen that it is possible to known a certain sentence p, even if we know its truth-conditions.

Let us focus on the intuitionistic revision proposed by Dummett [6] and Williamson [24]. Is their solution satisfactory? Marton [11, p. 86] observes that to answer this question, one should notice that any *verificationist* theory should include empirical propositions. So, Marton reformulates the question in the following way:

> Can Williamson's solution be extended to empirical propositions? This is certainly a highly problematic question, as Williamson repeatedly emphasized (e.g. 1994, 135–137), the intuitionistic approach to the paradox can only work if the intuitionistic semantics is also granted. No such generally accepted semantics of empirical propositions seems to be available, however.

This same fact was already pointed out by Prawitz [17] when he observed that the serious obstacles to the project of generalizing a verificationist theory to empirical discourse concern sentences for which there are no conclusive verifications (2002, p. 90). Thus, if knowability is an essential feature of the antirealist paradigm in philosophy, when applying antirealist theses to empirical sentences, things become at least complex. Mathematical truths are necessary, while empirical truths can be contingent and this is considered a problem for the antirealist thesis, since empirical sentences can hardly be proven conclusively, and sometimes not just *de*

---

[1] See [15], Sect. II.1.

[2] An interesting analysis of the issue can be found in Dalla Pozza [3].

[3] "My argument for the law of excluded middle and against the definition of "truth" in terms of "verifiability" is not that it is impossible to construct a system on this basis, but rather that it is possible to construct a system on the opposite basis, and that this wider system, which embraces unverifiable truths, is necessary for the interpretation of beliefs which none of us, if we were sincere, are prepared to abandon" (p. 682).

facto but because they are *intrinsically* unknowable.[4] Thus, it seems that the antirealist notion of truth cannot be easily associated with knowability in the case of empirical statements, since empirical sentences may be not decidable.[5]

A second problem for the antirealist concerns *undecidedness*: a stronger knowability paradox named *undecidedness paradox* is derivable from the intuitionistic revision. Percival [14] argues that the intuitionistic revision of the paradox involves a further paradox stating that there are no necessary undecided statements, which seems absurd also from the verificationist perspective. Consider the assumption that there are undecided statements in the intuitionistic and epistemic calculus:

(1) $\exists p(\neg Kp \wedge \neg K\neg p)$	Assumption(undecidedness)
(2) $(\neg Kp \wedge \neg K\neg p)$	From(1); instantiation
(3) $\forall p(\neg Kp \rightarrow \neg p)$	Intuitionistically equivalent to the denial of Non − Om
(4) $(\neg Kp \rightarrow \neg p)$	Instantiation of(3)
(5) $(\neg K\neg p \rightarrow \neg\neg p)$	Substitution of p with $\neg p$
(6) $\neg p \wedge \neg\neg p$	Contradiction from(2), (4)and(5)
(7) $\neg\exists p(\neg Kp \wedge \neg K\neg p)$	From(1)and(6)

In the above argument an intuitionistic contradiction follows. Thus, the antirealist using intuitionistic logic cannot hold that there are undecided statements and this seems absurd. A possible way to escape the conclusion is to use Williamson's strategy by formalizing undecidedness as:

$$\neg\forall p(Kp \vee K\neg p).$$

The above is classically, but not intuitionistically, equivalent to (1). So, it is only classically, but not intuitionistically, inconsistent with the result at line (6).

Has the logic of pragmatics LP some good points when handling the above problems?

[4] [18] points out that empirical and mathematical assertions can be justified by means of different grounds. He remarks that "a ground for the assertion of a numerical identity would be obtained by making a certain calculation, and outside of mathematics, a ground for asserting an observational sentence would be got by making an adequate observation". Dummett [5], in fact, points out that: "The intuitionist theory of meaning applies only to mathematical statements, whereas a justificationist theory is intended to apply to the language as a whole. The fundamental difference between the two lies in the fact that, whereas a means of deciding a range of mathematical statements, or any other effective mathematical procedure, if available at all, is permanently available, the opportunity to decide whether or not an empirical statement holds good may be lost: what can be effectively decided now will no longer be effectively decidable next year, nor, perhaps, next week" (p. 42).

[5] See also Hand [9].

5 An Outline of the Logic for Pragmatics LP

Dalla Pozza and Garola [4] proposed a pragmatic interpretation of intuitionistic propositional logic as a *logic of assertions*. They were mainly inspired by the logics of Frege and Dummett and by Austin's theory of illocutory acts.

Roughly speaking, the idea is to follow Frege distinguishing propositions from judgements. To briefly recapitulate Frege's distinction: the proposition has a truth value, while a judgement is the acknowledgement of the truth by a proposition. Propositions can be either true or false, while the act of judgement can be expressed through an act of assertion, which can be justified (hereafter "*J*") or unjustified (hereafter "*U*").

The idea of a pragmatic analysis of sentences/propositions has been developed by Reichenbach [19]. Following Frege and Reichenbach, in Dalla Pozza and Garola the assertion sign $\vdash$ consists of two parts: the horizontal stroke is a sign showing that the content is judgeable, the vertical stroke is a sign showing that the propositional content is asserted.[6] Differently from Frege's logical system, where assertive sentences cannot be nested, in Dalla Pozza and Garola's system pragmatic connectives are introduced to build complex formulas out of expressions of assertion.

Moreover, following Reichenbach's observations on assertions, i.e. that (i) assertions are part of the pragmatic aspects of language and (ii) assertions cannot be connected with truth-functional operators, in LP there are two sets of formulas: *radical* and *sentential* formulas. Every sentential formula contains at least a radical formula as a proper subformula. Radical formulas are semantically interpreted by assigning them with a (classical) truth value, while sentential formulas are pragmatically evaluated by assigning them a justification value (*J*, *U*), defined in terms of the intuitive notion of proof. Assertive connectives have a meaning which is explicated by the *BHK* (Brouwer, Heyting, Kolmogorov) intended interpretation of logical constants. Namely, atomic formulas are justified by a proof, while the justification of an implication is a method transforming a justification of the antecedent into a justification of the consequent, and so on.

The pragmatic language LP is the union of the set of radical formulas RAD and the set of sentential formulas SENT, which can be recursively defined:

$$\text{RAD}\gamma ::= p; \neg\gamma; \gamma_1 \wedge \gamma_2; \gamma_1 \vee \gamma_2; \gamma_1 \rightarrow \gamma_2; \gamma_1 \leftrightarrow \gamma_2.$$

$$\text{SENT (i) atomic assertive: } \eta ::= \vdash \gamma$$

$$\text{(ii) Assertive}\delta ::= \eta; \sim\delta; \delta_1 \cap \delta_2; \delta_1 \cup \delta_2; \delta_1 \supset \delta_2; \delta_1 \equiv \delta_2.$$

As proved by Dalla Pozza and Garola [4], classical logic is expressed in LP by means of those valid pragmatic assertions that are elementary (i.e. the sentential formulas that do not include pragmatic connectives). This classical fragment is called

---

6 From this perspective, notice that an assertion is a "purely logical entity" independent of the speaker's intentions and beliefs. For a different perspective see [10].

CLP). In this way, the corresponding radical formulas are tautological molecular expressions. On the other hand, intuitionistic logic is obtained by limiting the language of LP to complex formulas that are valid with atomic radical, even if the metalanguage is still classical. This intuitionistic fragment is called ILP.

The semantic rules for radical formulas are the usual Tarskian rules that specify the truth-conditions by means of a semantic assignment-function σ. Let γ_1, γ_2 be radical formulas, then:

(i) $\sigma(\neg\gamma_1) = 1$ iff $\sigma(\gamma_1) = 0$
(ii) $\sigma(\gamma_1 \wedge \gamma_2) = 1$ iff $\sigma(\gamma_1) = 1$ and $\sigma(\gamma_2) = 1$
(iii) $\sigma(\gamma_1 \vee \gamma_2) = 1$ iff $\sigma(\gamma_1 = 1)$ or $\sigma(\gamma_2) = 1$
(iv) $\sigma(\gamma_1 \rightarrow \gamma_2) = 1$ iff $\sigma(\gamma_1) = 0$ or $\sigma(\gamma_2) = 1$

There are also justification rules formalized by the pragmatic evaluation π governing the justification-conditions for assertive formulas in function of the σ assignments of truth-values for the radical atomic formulas (namely, π depends on the semantic function σ for radical atomic formulas). A pragmatic evaluation function is such that

$$\pi : \delta \in EN \longmapsto_{\pi} \delta \in \{J, U\}$$

Proposition 1 *Let γ be a radical formula. Then, $\pi(\vdash \gamma) = J$ iff there is a proof that γ is true, i.e. σ assigns to γ the value "true". Hence, $\pi(\vdash \gamma) = U$ iff no proof exists that γ is true.*

Proposition 2 *Let δ be a sentential formula. Then, $\pi(\sim \delta) = J$ iff a proof exists that δ is unjustified, namely that $\pi(\delta) = U$.*

Proposition 3 *Let δ_1, δ_2 be sentential formulas, then:*

- $\pi(\delta_1 \cap \delta_2) = J$ *iff* $\pi(\delta_1) = J$ *and* $(\delta_2) = J$
- $\pi(\delta_1 \cup \delta_2) = J$ *iff* $\pi(\delta_1) = J$ *or* $(\delta_2) = J$
- $\pi(\delta_1 \supset \delta_2) = J$ *iff a proof exists that* $\pi(\delta_2) = J$ *whenever* $(\delta_1) = J$
- $\pi(\delta_1 \equiv \delta_2)) = J$ *iff* $\pi(\delta_1 \supset \delta_2) = J$ *and* $\pi(\delta_2 \supset \delta_2) = J$

Proposition 4 *Let $\gamma \in RAD$. If $\pi(\vdash \gamma) = J$ then $\sigma(\gamma) = 1$*

Modus Ponens rule is provided for both (CLP) and ILP, respectively

$$\text{[MPP] If} \vdash \gamma_1, \vdash \gamma_1 \rightarrow \gamma_2 \text{ then} \vdash \gamma_2$$

and

$$\text{[MPP'] If } \delta_1, \delta_1 \supset \delta_2 \text{ then } \delta_2$$

where δ_1 and δ_2 contain only atomic radicals. Moreover, note that the justification rules do not always allow for the determination of the justification value of a complex sentential formula when all the justification values of its components are known. For instance, $\pi(\delta) = J$ implies $\pi(\sim \delta) = U$, while $\pi(\delta) = U$ does not necessary imply $\pi(\sim \delta) = J$.

In addition, a formula δ is pragmatically valid or p.valid (respectively invalid or p.invalid) if for every π and σ, the formula $\delta = J$ (respectively $\delta = U$). Note that if δ is p.valid, then $\sim\delta$ is p.invalid and if $\sim\delta$ is p.valid then δ is p.invalid. This is the criterion of validity for the pragmatic negation. We insert them just for completeness of exposition but we will not make use of them here, the same as for other pragmatic criteria of validity presented in [4].

Hence, no principle analogous to the truth-functionality principle for classical connectives holds for the pragmatic connectives in LP, since pragmatic connectives are partial functions of justification.

The set of radical formulas correspond to propositional formulas of classical logic, while the set of sentential formulas is obtained by applying the sign of assertion $\vdash$ to radical formulas. An assertion is justified by means of a proof and it cannot be iterated: so $\vdash\vdash \gamma$ is not a wff of LP. Nonetheless $\vdash \Box\gamma$, with $\Box$ in a S4 modality, is a wff of an extended pragmatic language with modal operators in the radical formulas. We will follow this suggestion when we will introduce the modal and epistemic operators in the intuitionistic fragment of LP. This fragment ILP is obtained limiting LP to complex formula valid with radical atomic formula. The axiom of ILP are:

$$
\begin{aligned}
&A1.\ \delta_1 \supset (\delta_2 \supset \delta_1)\\
&A2.\ (\delta_1 \supset \delta_2) \supset ((\delta_1 \supset (\delta_2 \supset \delta_3)) \supset (\delta_1 \supset \delta_2))\\
&A3.\ \delta_1 \supset (\delta_2 \supset (\delta_1 \cap \delta_2))\\
&A4.\ (\delta_1 \cap \delta_2) \supset \delta_1;\ (\delta_1 \cap \delta_2) \supset \delta_2\\
&A5.\ \delta_1 \supset (\delta_1 \cup \delta_2);\ \delta_2 \supset (\delta_1 \cup \delta_2)\\
&A6.\ (\delta_1 \supset \delta_3) \supset ((\delta_2 \supset \delta_3) \supset (\delta_1 \cup \delta_2) \supset \delta_3))\\
&A7.\ (\delta_1 \supset \delta_2) \supset ((\delta_1 \supset (\sim \delta_2)) \supset (\sim \delta_1))\\
&A8.\ \delta_1 \supset ((\sim \delta_1) \supset \delta_2)
\end{aligned}
$$

The assertion sign is not a predicate and asserted sentences cannot be embedded, for instance, in the antecedent of an implication. As observed, this is a classical feature of assertion and it is what Geach [8] calls *Frege's point*. Moreover, an assertion sign cannot be within the scope of a classical (truth conditional) connective, since it works in what is called "pragmatic capacity" [19].

Sentential formulas have an intuitionistic-like behaviour and can be translated into modal system S4, where $\vdash \gamma$ can be translated as $\Box\gamma$, meaning that "there is an (intuitive) proof of the truth of γ" in the sense of empirical or logical procedures of proof.

Briefly put, the modal meaning of pragmatic assertions is provided by the following semantic translation of pragmatic connectives. Sentential formulas can be translated into the classical modal system S4 as in the following table:

$(\vdash \gamma) \,/\, \Box\gamma$
$\sim\vdash \gamma \,/\, \Box\neg\Box\gamma$
$(\vdash \gamma_1 \cap \vdash \gamma_2) \,/\, \Box(\gamma_1) \wedge \Box(\gamma_2)$
$(\vdash \gamma_1 \cup \vdash \gamma_2) \,/\, \Box(\gamma_1) \vee \Box(\gamma_2)$
$(\vdash \gamma_1 \supset \vdash \gamma_2) \,/\Box(\Box(\gamma_1) \rightarrow \Box(\gamma_2))$

Classical and intuitionistic formulas are related by means of the following "bridge principles":

(a) $\vdash (\neg\gamma) \supset \sim\vdash (\gamma)$
(b) $(\vdash \gamma_1 \cap \vdash \gamma_2) \equiv \vdash (\gamma_1 \wedge \gamma_2)$
(c) $(\vdash \gamma_1 \cup \vdash \gamma_2) \supset \vdash (\gamma_1 \vee \gamma_2)$
(d) $(\vdash \gamma_1 \rightarrow \gamma_2) \supset (\vdash \gamma_1 \supset \vdash \gamma_2)$

The formula (a) states that from the assertion of not-γ, the non-assertability of γ can be inferred. (b) states that the conjunction of two assertions is equivalent to the assertion of a conjunction; (c) states that from the disjunction of two assertions one can infer the assertion of a disjunction. Finally (d) expresses the idea that from the assertion of a classical material implication follows the pragmatic implication between two assertions. Note that such principles hold in an extension of ILP with classical connectives. We name such fragment ILP$^+$.

6 A Pragmatic Treatment of the Knowability Paradox

Let us present the *Knowability Paradox* in the framework of ILP enriched with a knowledge operator K and aletheic modality. Notice that such a logic cannot be ILP or ILP$^+$ because, as mentioned, intuitionistic logic is obtained by limiting the language of LP to complex formulas that are valid with atomic radical, even if the metalanguage is still classical. Given the above characterization of ILP, the formula $\vdash \Diamond Kp$ is not a wff of ILP. We extend ILP with a knowledge operator K and alethic modality. Concerning modality: we have already observed that $\vdash \Box\gamma$, with $\Box$ in an S4 modality, is a wff of an extended pragmatic language with modal operators in the radical formulas. Regarding the knowledge operator K: it is possible to treat it using some analogous *invariance principles* given by Ranalter in [20] for the *ought* operator.[7] Moreover, for the sake of simplicity we will not make use of quantifiers. We start with a suitable formulation of the Knowability Principle in KILP:

[7] A similar *intermediate logic* has been developed in [1].

$$(\text{KP'}) : (\vdash p \supset \vdash \Diamond Kp). \qquad \text{(instance of knowability in KILP)}$$

(KP') is a wff of KILP and states that there exists a method transforming a proof of p into a proof of the possibility of knowledge that p, which is a stronger claim with respect to (KP), i.e. "for every p, $p \rightarrow \Diamond K\ p$". In (KP') one claims that there is a proof of the knowability of p. The principle of non-omniscience in KILP is—again—stronger than (Non-Om), namely:

$$(\text{Non-Om'}): \vdash p \cap \sim\vdash Kp \ \text{(instance of Non-Omniscience in KILP)}.$$

Non-Om' states that there is a proof of p without knowing to know that p. If so, Non-Om' says something different form the fact it should express: i.e. non-omniscience.

Observe that the arguments leading to the knowability paradox cannot be formulated in KILP, first of all for syntactic reasons. Let us consider the first argument:

$$(1') \vdash p \cap \sim\vdash Kp$$

the substitution of "p" with "$p \wedge \neg Kp$" cannot be executed, since formulas with classical connectives are not wff of KILP. Again, the substitution of the radical formula "p" with "$\vdash p \cap \sim\vdash Kp$" in KP' does not work, since the sign of assertion cannot be nested. Moreover, from the substitution "$\vdash p$" with "$\vdash p \cap \sim\vdash Kp$" in KP', it merely follows:

$$(2') \ (\vdash p \cap \sim\vdash Kp) \supset \vdash \Diamond Kp$$
$$(3') \vdash \Diamond Kp \qquad \textit{modus ponens} \text{ from } (1') \text{ and } (2').$$

Let us now consider the second independent argument of the paradox. It is worth noting that it is impossible to state the assumption for the *reductio* in KILP; namely both

$$(4^*) \ \vdash K\ (p \wedge \neg Kp)$$
$$(4^{**}) \vdash K(\vdash p \cap \sim\vdash Kp)$$

are not wff of KILP, since (4*) contains classical connectives, while in (4**) the sign of assertion is nested. Moreover, consider a semantic reading of (4*): there is a proof that we know that p is true and that we do not know that p is true. It does not make any sense! Hence, there is no way to reproduce the paradox in the language of KILP. Consequently, the argument leading to the paradox is stopped at the early inferential steps.

7 A Comparison with the Intuitionistic Solution

One could argue that the result just obtained in KILP is not surprising if KILP is an adequate extended fragment of intuitionistic logic. We have argued that in the intuitionistic solution KP can be weakened and formulated as a valid intuitionistic formula. Does KILP has any chances to supersede the antirealist difficulties skected in the paper, in Sect. 4? First, consider a preliminary remark. Observe that, differently from intuitionism, in KILP:

(A) A is true if and only if it is possible to exibit a direct justification for A

does not hold.

Indeed, for an antirealist truth is epistemically constrained, while subscribers of LP hold that what can be properly justified in LP are (assertive) acts, and propositions can be true or false. Notice, moreover, that the use of logical constants in the metalanguage of KILP is classical. That is why (A) is false in LP. In LP we have to distinguish a semantic and a pragmatic level. From the fact that a certain sentence is true it does not mean that the same sentence is justified. If (A) is false in LP then KP does not follow. In fact, KP is the result of:

(A) A is true if and only if it is possible to exibit a direct justification for A.

and

(B) If it is possible to exhibit a direct justification for A, then it is possible to know that A.

As already been mentioned, in putting (A) and (B) together, we get the knowability principle:

(KP) If A is true, then it is possible to know that A.

This result is in accordance with the syntactic translation given above: In KP' we have observed that we have a proof of the knowability of p whereas in KP we just claim its knowability. If KP' holds then KP holds but not *vice versa*.

Consider what happens in KILP with undecidedness.

First, notice that the argument leading to the paradox of undecidedness cannot be replicated in KILP, since we cannot even express an instance of undecidedness: Merely from a syntactical point of view

$$(\sim\vdash Kp \cap \sim\vdash K\neg p)$$

is not, in fact, a wff of KILP. A slightly different notion of undecidedness can be expressed by the formula:

$$\textit{there is a } p \textit{ such that} \sim (\vdash Kp \cup \sim\vdash Kp)$$

namely, there is a p such that it is not provable that the assertion of Kp holds or that the assertion Kp does not hold. Let us consider the formula

$$\textit{there is a } p \textit{ such that } (\vdash p \cap \sim\vdash Kp)$$

which expresses non-omniscience in KILP. An instance of the denial of non-omniscience can be now expressed in the following way:

$$(0) \sim (\vdash p \cap \sim\vdash Kp) \text{ Negation of an instance of (Non-Om').}$$

Observe that, also with the above version of undecidenss plus Non-Om' the argument leading to the *Undecidedness Paradox of Knowability* cannot be expressed in KILP. In fact, let us consider the following steps:

(1) $\sim (\vdash Kp \cup \sim\vdash Kp)$	Assumption (instance of undecidedness)
(2) $(\sim\vdash Kp \supset \sim\vdash p)$	Equivalent to (0)
(3*) $(\sim\vdash K\neg p \supset \sim\vdash \neg p)$	Substitution of "p" with "$\neg p$" not allowed in KILP

(1) can be assumed in order to express a stronger form of undecidability, understood as a the existence of a proof of the impossibility of obtaining decidability. Notice that the negation of the excluded middle is a contraddiction in intuitionistic logic, whereas the justification value of (1) might be undeterminate in KILP, according to the justification rules of ILP expanded to KILP. Indeed, there is a formal equivalence only among theorems of intuisionistic logic and the corresponding p.valid formula of ILP (expanded in the obvious way to KILP), while it does not follow for formulas different from theorems.

(2) can be derived and a reading of (2)—suggested by the BHK interpretation of logical constants—is the following one: there is a method which transforms a proof that Kp cannot be proven into a proof that p cannot be proven. While a classical reading of (2), namely $\neg Kp \rightarrow \neg p$, means that ignorance entails falsity, the pragmatic reading of (2) deals with the conditions of provability of K. Finally, (3*) cannot be obtained in KILP, since it is not possible to substitute "p" with "$\neg p$" (the negation is classical).

Perhaps, if one wants to express undecidedness by means of conjunction as in the original paradox, the following might do. Consider the undecidedness paradox of knowability expressed in an extension of KILP with classical negation. We name it KILP$^+$. In KILP$^+$, undecidedness can be expressed with

$$\textit{there is a } p \textit{ such that } (\sim\vdash Kp \cap \sim\vdash K\neg p).$$

Consider the following steps:

(1)'	$\sim\vdash Kp \cap \sim\vdash K\neg p$	Assumption(instance of undecidedness)
(2)'	$(\sim\vdash Kp \supset\sim\vdash p)$	Equivalent to (0)
(3)'	$(\sim\vdash K\neg p \supset\sim\vdash \neg p)$	Substitution of " p" with "$\neg p$" allowed in KILP$^+$
(4)'	$\sim\vdash p \cap \sim\vdash \neg p$	Application of the conjuncts of(1)$'$to(2)$'$and(3)'

Notice that (4)' does not involve a paradoxical consequence. The fact that we do not have a proof of p, but also we do not have a proof of $\neg p$ is rather common for empirical sentences which are not decidable.

Unlike the treatment of the undecidedness paradox of knowability in intuitionistic logic, KILP in its extension KILP$^+$ does not involve the denial of undecided sentences. So one could argue either that the paradox is not formalizable in KILP or that it is not paradoxical in one extension (KILP$^+$) of it. This seems to be an advantage of KILP over intuitionistic logic.

8 Conclusions

In this paper paper we have analysed the paradox of knowability asking if it is reproducible within a logic for pragmatics (LP), especially in an extension of an intuitionistic fragment of it, KILP. We have shown the strict limits of the proposal, but also some advantages: the most important one concerns undecideness of contingent sentences.

Notice that the negation of a sentence in intuitionistic logic means that the proposition implies the absurd and this makes sense in mathematics, while—pretheoretically—the negation of a contingent empirical proposition does not imply the absurd. On the contrary, the pragmatic negation means that there is a proof that a certain proposition is not (or cannot be) proved. The formal behaviour of the pragmatic negation can be properly understood when one take into consideration the excluded middle. It can be written as $(\vdash p \cup \sim\vdash p)$. p is an atomic formula atomic and it allows only an empirical procedure of proof; the following situation is possible: we do not have an empirical proof procedure for asserting p and we do not have any empirical procedure of proof for not asserting p. Therefore, $(\vdash p \cup \sim\vdash p)$ is not justified (see Proposition 3).

This property of the pragmatic negation combined with the possibility to express empirical procedures of proof in the language of LP shows some possible advantages with respect to intuitionistic logic when dealing with empirical sentences.

References

1. Bellin, G., & Biasi, C. (2004). Towards a logic for pragmatics: Assertions and conjectures. *Journal of Logic and Computation*, *14*(4), 473–506.
2. Church, A. (2009). Referee reports on Fitch's "A definition of value". In J. Salerno (Ed.), *New essays on the knowability paradox* (pp. 13–20). Oxford: Oxford University Press.
3. Dalla Pozza, C. (2008). Il problema della demarcazione. Verificabilità, falsificabilità e conferma bayesiana a confronto. Ese, Lecce.
4. Dalla Pozza, C., & Garola, C. (1995). A pragmatic interpretation of intuitionistic propositional logic. *Erkenntnis*, *43*(1), 81–109.
5. Dummett, M. (2004). *Truth and the past.* New York: Columbia University Press.
6. Dummett, M. (2009). Fitch's paradox of knowability. In J. Salerno (Ed.), *New essays on the knowability paradox* (pp. 51–2). Oxford: Oxford University Press.
7. Fitch, F. (1963). A logical analysis of some value concepts. *Journal of Symbolic Logic*, *28*(2), 135–42.
8. Geach, P. (1965). Assertion. *Philosophical Review*, *74*(4), 449–65.
9. Hand, M. (2010). Antirealism and universal knowledge. *Synthese*, *173*(1), 25–39.
10. Lackey, J. (2007). *Norms of assertion. Noûs*, *41*(4), 594–626.
11. Marton, P. (2006). Verificationists versus Realists: The battle over knowability. *Synthese*, *151*, 81–98.
12. Murzi, J. (2010). Knowability and bivalence: Intuitionistic solutions to the paradox of knowability. *Philosophical Studies*, *149*(2), 269–81.
13. Pap, A. (1962). *An introduction to the philosophy of science*. New York: Free Press.
14. Percival, P. (1990). Fitch and intuistionistic knowability. *Analysis*, *50*(3), 182–7.
15. Poincaré, E. (1914). *Science and Method Analysis.* London, New York: T. Nelson and Sons.
16. Prawitz, D. (1987). Dummett on a theory of meaning and its impact on logic. In B. M. Taylor (Ed.), *Contributions to philosophy* (pp. 117–65). Dordrecht: Martinus Nijhoff.
17. Prawitz, D. (2002). Problems for a generalization of a verificationist theory of meaning. *Topoi*, *21*(1–2), 87–92.
18. Prawitz, D. (2012). The epistemic significance of valid inference. *Synthese*, *187*, 887–898.
19. Reichenbach, H. (1947). *Elements of symbolic logic*. New York: Free Press.
20. Ranatler, K. (2008). A semantic analysis of a logic for pragmatics with assertions, obligations, and causal implication. *Fundamenta Informaticae*, *84*(3–4), 443–470.
21. Salerno, J. (Ed.). (2009). *New essays on the knowability paradox*. Oxford: Oxford University Press.
22. Schilpp, P. A. (Ed.). (1951). *The philosophy of Bertrand Russell* (3rd ed.). New York: Tudor Publ. Co.
23. Sundholm, G. (1997). Implicit epistemic aspects of constructive logic. *Journal of Logic, Language and Information*, *6*(2), 191–212.
24. Williamson, T. (1982). Intuitionism disproved. *Analysis*, *42*(4), 203–7.
25. Williamson, T. (1994). Never say never. *Topoi*, *13*(2), 135–145.

A Dialetheic Interpretation of Classical Logic

Massimiliano Carrara and Enrico Martino

Abstract According to classical logic, the acceptance of a *dialetheia*, a proposition that is both true and false, entails *trivialism* the output that every sentence is true. One way to accept dialetheias but avoid trivialism is to reject the general validity of classical logic, which is the view held by dialetheists, supporters of the existence of dialetheias. In *The Logic of Paradox* (LP), Priest adopts the *material conditional*, identifying $A \rightarrow B$ with $\neg A \vee B$. He argues that this is not a genuine conditional because it invalidates *modus ponens* (MP), an essential rule governing the use of the conditional. In subsequent works he introduces a genuine conditional and tries to avoid Curry's paradox by invoking a highly problematic modal semantics. The aim of our paper is to argue that a dialetheist can stick to the material conditional and recover the whole of classical logic without falling into trivialism. Our strategy sets forth a way of understanding the notion of assumption suitable for the dialetheic perspective. We show the inadequacy of formal classical logic to capture the intended exclusivity of negation. Finally, we argue that the material conditional is adequate to provide a dialetheic solution to semantic paradoxes.

Keywords Dialetheism · Classical logic · Curry's Paradox

1 Introduction

Dialetheism holds the thesis that there exist *dialetheias* (i.e., propositions that are both true and false). Priest (for example, in Priest [4–6]) claims that dialetheism supplies the best solution to all the self-reference paradoxes. The paradigmatic example of a

M. Carrara (✉) · E. Martino
FISPPA Department, University of Padua, Padua, Italy
e-mail: massimiliano.carrara@unipd.it

E. Martino
e-mail: enrico.martino@unipd.it

R. Ciuni et al. (eds.), *Recent Trends in Philosophical Logic*,
Trends in Logic 41, DOI: 10.1007/978-3-319-06080-4_4,

self-reference paradox is the strengthened liar paradox, having the form

$$(a) : (a) \text{ is not true},$$

which is solved, according to Priest, by holding that (a) is both true and not true.

In classical logic, the presence of a *dialetheia* entails *trivialism*—the truth of all sentences—through the classical rule of *ex contradictione quodlibet* (ECQ). Dialetheism avoids trivialism by rejecting the general validity of classical logic. We want to argue that a dialetheist can accept the general validity of classical logic without falling into trivialism. We do it by refining the notion of assumption and by exploiting a relevant feature of the material conditional.

2 The *Dialetheic* Meaning of Logical Constants

Priest maintains that if dialetheism is to be a tenable view, the dialetheic meaning of logical constants must be adopted even in the meta-language:

(MLC) *The meaning of logical constants is to be the same in the object language and in the meta-language.*

Consider the language L of first-order logic. According to MLC, the meaning of logical constants can be translated from the meta-language into the object language by means of the usual homophonic Tarskian clauses, where the metalinguistic logical constants are understood *dialetheically*. The clauses are the following:

1. $\neg A$ is true if$_{df}$ A is not true.
2. $A \wedge B$ is true if$_{df}$ A is true and B is true; similarly for the universal quantifier ($\forall$).
3. $A \vee B$ is true if$_{df}$ A is true or B is true; similarly for the existential quantifier ($\exists$).
4. $A \rightarrow B$ is true if$_{df}$ A is not true or B is true.

Clause (4) defines the so-called material conditional, which was adopted by Priest in *The Logic of Paradox* (LP) [4]. In subsequent works (for example, [5]), Priest introduces other kinds of highly problematic conditionals (see last section). However, we are here mainly interested in the language of LP.

What is the *dialetheic meaning* of the logical constants? The dialetheic meaning of conjunction, disjunction and quantifiers seems to be the usual classical one. What seems to be highly problematic, however, is how to understand *dialetheic negation*. The understanding of a dialetheia seems to presuppose a non exclusive meaning of negation, suitable for making a sentence compatible with its negation, in contrast to the classical (and intuitionistic) meaning. Priest denies, however, that the acceptance of dialetheias depends on the adoption of some anomalous negation. In his discussion of *Boolean negation* (i.e. exclusive negation), [6, Part II, Sect. 5] argues that no negation can rule out, by virtue of its very meaning, the existence of dialetheias. According to his view, this would be a "surplus" that no understanding of negation can preclude. He maintains that every attempt to defend the existence of Boolean negation has been a *petitio principii* and draws the following conclusion:

Boolean negation negated (BNN) *The dialetheist is at liberty to maintain that Boolean negation has no coherent sense* [5, p. 98].

BNN is of crucial importance for the dialetheic solution to semantic paradoxes. In fact, if a dialetheist recognized a legitimate sense to Boolean negation, he could use it in the meta-langage and translate it into the object language through clause (1) mentioned above. In that case, the paradoxes would be reproduced, and the dialetheic solution would fail.

We think that BNN—as a general claim about the use of negation in natural language—is untenable: no one can avoid the use of Boolean negation in natural language. Priest himself, when claiming that certain classical rules are not valid, must use an exclusive "not", on pain of nullifying his own claim. The thesis itself (that an exclusive negation does not exist) exploits the notion of *exclusivity*. If this notion is defined in terms of the dialetheic negation, it doesn't reach the intended meaning and the distintion between exclusive and non-exclusive negation collapses. On the other hand, if exclusivity is taken as primitive in the intended sense, then the exclusive negation can be defined in the object language through the clause:

(1*) $\neg A$ is true if$_{df}$ the truth of A is excluded.

Anyway, we concede, for the sake of argument, that the meaning of negation, both in the object language and in the meta-language, is the dialetheic one (i.e., compatible with the existence of true contradictions, both linguistic and meta-linguistic).

Let us consider the semantics of LP. Let L be the language of first-order logic. A dialetheic interpretation of L consists of a pair (D, a) where D is a non-empty domain of individuals and a is an assignment (intended as a one-many relation) that assigns to every atomic sentence one or both the truth-values 1, 0. Suppose, for the sake of simplicity, that L has a name for every member of D. We say that an atomic sentence is true (is false) if it is assigned the value 1 (the value 0). The truth-values of a compound sentence are inductively defined as follows:

(a) $\neg A$ is true if$_{df}$ A is false; it is false if$_{df}$ A is true.
(b) $A \wedge B$ is true if$_{df}$ A is true and B is true; it is false if$_{df}$ A is false or B is false; similarly for the universal quantifier ($\forall$).
(c) $A \vee B$ is true if$_{df}$ A is true or B is true; it is false if$_{df}$ A is false and B is false; similarly for the existential quantifier ($\exists$).
(d) $A \rightarrow B$ is true if$_{df}$ A is false or B is true; it is false if$_{df}$ A is true and B is false.

The adoption of this semantics, together with MLC, suggests that clauses (1)–(4) should be equivalent to clauses (a)–(d). So we assume that the meta-linguistic negation used in clause (1) is understood in such a way that clauses (1) and (a) characterize the same notion of negation in the object language. That is in agreement with Priest's inter-definability of falsity and negation [6, p. 46]:

$$T(\ulcorner\neg A\urcorner) \leftrightarrow F(\ulcorner A\urcorner),\ F(\ulcorner\neg A\urcorner) \leftrightarrow T(\ulcorner A\urcorner)$$

(where T and F are the truth and the falsity predicate respectively).

Therefore, we consider clauses (1)–(4) and clauses (a)–(d) as equivalent. It follows that falsity and untruth are identified. In formal semantics, this means that the fact that an assignment assigns to an atomic sentence A the value 0 is to be understood as a convention to the effect that A represents an untrue (possibly also true) proposition. In other words, to say that A is untrue does not mean that the assignment does not assign to A the value 1, but that it assigns to A the value 0.

Priest actually seems to be puzzled about the identification of untruth and falsity. He observes that the principle $F(\ulcorner A \urcorner) \rightarrow \neg T(\ulcorner A \urcorner)$ could, at least *prima facie*, be rejected on the ground that the falsity of A does not exclude its truth. We think, however, that dialetheists should not worry about that since, according to their view, negation is nonexclusive: $\neg T(\ulcorner A \urcorner)$ is compatible with $T(\ulcorner A \urcorner)$. This distinction between falsity and untruth seems to be appropriate to gap-theorists: for them an untrue sentence may be neither true nor false. But, since for the dialetheist any proposition is true or false, it is hard to see what can distinguish a false proposition from an untrue one.

Moreover, consider the formalization of the *strengthened liar paradox*. Let $k = \ulcorner \neg T(\underline{k}) \urcorner$. The paradox is solved by assigning to the atomic sentence $T(\underline{k})$ the values 1 and 0. So, according to clause (a), $\neg T(\underline{k})$ is false; and in order to read $\neg T(\underline{k})$ as "$T(\underline{k})$ is untrue", falsity and untruth must be identified. In other words, for the dialetheist there is no room for distinguishing the strengthened liar from the simple liar.

For these reasons we consider the following principles dialetheically valid:

(2*) $F(\ulcorner A \urcorner) \leftrightarrow T(\ulcorner \neg A \urcorner) \leftrightarrow \neg T(\ulcorner A \urcorner)$

Priest observes that, when dealing with dialetheias, certain rules of classical logic are invalid. For instance, the material conditional invalidates *modus ponens* (MP). Indeed, if A is a dialetheia, A and $\neg A \vee B$ (i.e, $A \rightarrow B$) are true for any B, so that, by MP, B would follow. For this reason he claims that MP is *quasi-valid*. In general, he calls *quasi-valid* the classical rules that are valid as far as no dialetheia is involved.

One might observe, as Priest does, that the material conditional is not a genuine conditional because "any conditional worth its salt should satisfy the *modus ponens* principle". However, we will show that, in a dialetheically intelligible sense, the material conditional satisfies all classical inference rules, so it can be considered a genuine conditional.

3 Material Validity

Consider the general notion of logical validity:

An inference rule is valid if it preserves truth from the premises to the conclusion.

However, the definition of truth-preservation involves, in turn, a meta-linguistic conditional. An inference rule is truth-preserving if it satisfies the following condition:

If the premises are true *so* is the conclusion.

Let us adopt the material conditional even in the meta-language. Since such conditional is used in LP, that is in agreement with MLC. So, the truth preservation of MP is expressed as follows:

(3.1) A is false or $(\neg A \vee B)$ is false or B is true.

(3.1) is satisfied even if A is a dialetheia because in this case A is false. Observe that if A is a dialetheia and B is false, then (3.1) is a meta-linguistic dialetheia: since the first disjunct is true, the disjunction is true; and since all three disjuncts are false, the disjunction is false. Despite some counterexamples to the truth preservation of MP, a dialetheist can nevertheless conclude that MP is *always* truth-preserving. In other words, the meta-linguistic proposition "the material conditional satisfies MP" is always true; sometimes it is a dialetheia. Thus, the dialetheist cannot *reject* the validity of MP for the material conditional. In general, we will show that, by adopting the material conditional in the meta-language, dialetheists can accept the entirety of classical logic, maintaining their solution to the paradoxes and avoiding trivialism.

Let ND be the classical system of natural deduction for first-order logic. We want to prove the following theorem:

Material validity (MV) *For every formal proof of ND, either the conclusion is true or some assumption is false.*

Let $I = (D, a)$ be a dialetheic interpretation. A classical interpretation $I' = (D, a')$ is a *sub-interpretation* of I if, for every sentence A, the I'-value of A is one of the I-values of A. Since the above clauses (a)–(d) hold both classically and dialetheically, a classical sub-interpretation of I is obtained through any assignment a' that, for every atomic sentence A, chooses one of the a-values of A.

Lemma 1 *Let I be a dialetheic interpretation for first-order logic, and let I' be any classical sub-interpretation of I. If a formula A is only I-true (only I-false), then it is I'-true (I'-false).*

Proof By an easy induction on the complexity of A. □

Proof (of MV) By way of reduction, let p be a proof of B from the assumptions $A_1, \ldots, A_n$, and let I be a dialetheic interpretation that makes $A_1, \ldots, A_n$ only true and B only false. By the lemma, any classical sub-interpretation of I makes $A_1, \ldots, A_n$ true and B false, against the classical soundness theorem. □

One may wonder if this proof is dialetheically correct. Indeed, the proof uses the notion of *only true* and proceeds by reduction to absurdity. Now, a well-known criticism to dialetheism concerns the difficulty of expressing the notion of exclusive truth dialetheically. In fact, *only true* is usually understood as "true and not false" with the exclusive *not*, which fails to be available to a dialetheist. However, a dialetheist needs the possibility of expressing such a notion. For instance, having held that a sentence may be true, false or both, the dialetheist should be able to reason by cases to distinguish three possible cases: *true only*, *false only*, *true and false*.

Let us consider two of Priest's arguments for defending the possibility of expressing dialetheically the notion of *true only*. The first is simply that a dialetheist can express the fact that a sentence is only true by using these very words, just as a classicist does. According to this claim, *only true* should be defined as "true and not false" even dialetheically. In fact, as already observed, Priest maintains that dialetheism does not alter the classical meaning of negation. The difference between a classicist and a dialetheist would be solely that the first pretends to guarantee consistency but the second does not:

> What the dialetheist cannot do, whether the topic is paradox or anything else, is ensure that views expressed are consistent. The problem then, for a dialetheist—if it is a problem—is that they can say nothing that *forces* consistency. But once the matter is put this way, it is clear that a classical logician cannot do this either. Maybe they would like to; but that does not mean they succeed. Maybe they intend to; but intentions are not guaranteed fulfilment. Indeed, it may be logically impossible to fulfil them, as, for example, when I intend to square the circle [6, p. 106].

We think that this argument misses the point. Expressivity has nothing to do with warranty. The problem is not how to say something that forces consistency; it is rather how to say *what consistency is*. The comparison to the problem of squaring the circle is misleading. Concerning that problem, it is not in question what squaring the circle means. On the contrary, the problem we are dealing with is what it means for a dialetheist to say that a sentence A is *only true*. The answer that it simply means that A is true and not false makes *"only true"* collapse into "true *tout court*". Since from $T(\ulcorner A \urcorner)$ it follows $\neg\neg T(\ulcorner A \urcorner)$ and, by (2*), $\neg F(\ulcorner A \urcorner)$, *any* true sentence is true and not false.

Priest's second argument is more plausible. He tries to make up for the lack of *exclusive negation* by introducing the notion of *rejection* of a proposition, to be clearly distinguished from the *acceptance* of negation ([5, 7]; for a general introduction to the topic see [8]). Acceptance and rejection are cognitive states corresponding to the linguistic *acts* of asserting and denying respectively. According to Priest, while one can accept both a proposition and its negation, one cannot accept and reject the same proposition. Thus, when recognizing that a proposition is a dialetheia, dialetheists cannot reject either it or its negation. Besides, when rejecting a proposition, they must accept its negation, excluding it from being a dialetheia. No connective, by virtue of its logical meaning, can serve the purpose of expressing the exclusion of the truth (or the falsity) of A, but one can express that by denying A (or $\neg A$).

Although Priest does not say anything, as far as we know, about the notion of *assumption*, we think that this can be treated, in turn, as an *illocutory act*. We propose to legitimate the possibility of assuming, in the course of a proof, that a sentence is *only* true (*only* false), where such assumptions are to be dialetheically understood as illocutory acts. Just as people can assert or deny something, they can also assume it. Furthermore, they can assume or assert something in an *exclusive mode*. In this vein, even if dialetheists reject exclusivity as part of the meaning of a proposition, they can accept it as a mode of an illocutory act. In this way, in accordance with the claim that a proposition may be true, false or both, a dialetheist can recover the possibility of distinguishing the three possible cases assuming, in turn, that it is (i) only true,

(ii) only false or (iii) both true and false. In this sense we think that the use of the notion of *truth only* in MV is dialetheically acceptable.

As to the procedure by reduction to absurdity, dialetheists can recognize that all classical inference rules are truth preserving under the assumption that all atomic sentences are only true or only false. In this way they can recover the classical soundness theorem. Similarly, they can rightfully suppose that the assumptions of a classical proof are only true and that the conclusion is only false and then reject such supposition in virtue of the soundness theorem. For these reasons we conclude that the proof of MV is dialetheically acceptable.

Of course, one can object that, according to dialetheism, material validity is not genuine validity because the truth of the assumptions does not guarantee the truth of the conclusion. However, material validity is adequate to formalize hypothetical reasoning even from a dialetheic perspective. When reasoning under certain assumptions, mathematicians do not generally know whether their assumptions are *actually* true. So what even a classical mathematician can know from a formal proof is that the conclusion is true or some assumption is false. Indeed, a classicist who knows that the assumptions are true can—unlike the dialetheist—assert the conclusion. However, a dialetheist can explain this fact by holding that, when the classicists know that the assumptions are true, they erroneously believe themselves to be in a position of rejecting the negations of the assumptions. And such a rejection is the appropriate ground for asserting the conclusion. In fact, even dialetheists, when they are in the position of rejecting the negations of all assumptions, can assert the conclusion.

By virtue of the possibility of assuming that a proposition is only true, a dialetheist may reformulate the meta-theorem MV as follows:

(MV*) *Given any dialetheic interpretation of L, a classical proof leads to a true conclusion under the assumption that the hypotheses are only true.*

Priest himself tries to recapture classical logic by introducing the notion of a quasi-valid inference rule (see in Priest [5, Sect. 8.5]). However, he puts the question in terms of extra-logical notions such as that of *rational acceptability* and *rational rejectability*.

An example is the way he explains the quasi-validity of the disjunctive syllogism (DS). He introduces the following principle about rational rejection:

(Principle R) *If a disjunction is rationally acceptable and one of the disjuncts is rationally rejectable, then the other is rationally acceptable.*

Then he justifies DS as follows:

> Suppose that $A \wedge (\neg A \vee B)$ is rationally acceptable. This entails $(A \wedge \neg A) \vee B$, which is therefore rationally acceptable. But provided $A \wedge \neg A$ is rationally rejectable (as it often will be...), then, by principle R, B is rationally acceptable. In other words, it is reasonable to accept the conclusion of a DS argument provided the contradiction involved is reasonably rejectable [5, p. 114].

The drawback is that since it is generally not decidable whether $A \wedge \neg A$ is rejectable, it is not decidable whether an application of a DS argument is correct or not.

In contrast, in our approach the dialetheic justification of DS is that B is true under the assumptions that $A \wedge (\neg A \vee B)$ is true and $A \wedge \neg A$ is *only* false. Observe that this formulation is purely logical since the act of assuming (possibly in exclusive mode) is essential to logical reasoning. Besides, logic allows us to assume *any* proposition, quite independently of its plausibility.

4 Classical Logic and Boolean Negation

We believe, unlike dialetheists, that Boolean negation is essential for thought and reasoning. But (MV*) is a meta-logical result of some interest even for a classical logician. It shows a remarkable limitation of formal logic. The classical inference rules are inadequate to capture the intended exclusivity of negation. In fact, as we have seen, any formal proof can be interpreted in a dialetheic model (with a non exclusive negation) under the assumption that all hypotheses are *only* true. In other words, the use of Boolean negation can be confined to the meta-language. One may compare this inadequacy with that concerning the notion of finiteness. Formal Peano arithmetic is inadequate for capturing the notion of finiteness even though this is essential for grasping the concept of a natural number. Similarly, formal logic is inadequate for capturing the notion of exclusive negation even though this is essential for reasoning and, more generally, for the understanding of natural language.

5 Semantic Paradoxes

The semantic paradoxes can be solved in our framework by holding that, although all instances of Tarski's schema are true, some of them are nevertheless dialetheias.

Consider, for example, Curry's paradox. The paradox is derived in natural language by proving sentences such as the following:

(b): If sentence (b) is true, then Santa Claus exists.

Proof Suppose that the antecedent of the conditional in (b) is true. This means that sentence (b) is true. Then, by MP, Santa Claus exists. So, we have proved the consequent of (b) under the assumption of its antecedent. In other words, we have proved (b). It follows, by MP, that Santa Claus exists. □

Of course, we could replace "Santa Claus exists" with any arbitrary sentence. As a result, every sentence can be proved, and trivialism follows.

Let us formalize the paradox in the language of first-order arithmetic extended with a truth predicate T satisfying Tarski's schema:

$$T(\ulcorner A \urcorner) \leftrightarrow A.$$

Given any sentence A, by means of the usual method of diagonalization, one can find a number k such that

$$k = \ulcorner T(\underline{k}) \rightarrow A \urcorner.$$

We can derive A as follows:

1	(1)	$T(\underline{k}) \leftrightarrow (T(\underline{k}) \rightarrow A)$	Tarski's schema
2	(2)	$T(\underline{k})$	Assumption
1, 2	(3)	$T(\underline{k}) \rightarrow A$	2 (MP)
1, 2	(4)	A	2, 3 (MP)
1	(5)	$T(\underline{k}) \rightarrow A$	2, 4 (I→)
1	(6)	$T(\underline{k})$	1, 5 (MP)
1	(7)	A	5, 6 (MP)

According to MV, a dialetheist can conclude that either (7) is true or (1) is false. Thus the dialetheist can escape trivialism by holding that some instances of Tarski's scheme are dialetheias. Observe that *trivialism* follows if, instead of using Tarski's schema in the biconditional form

$$T(\ulcorner A \urcorner) \leftrightarrow A$$

we adopt the following inference rules:

$$(*)\frac{T\ulcorner A \urcorner}{A} \qquad (**)\frac{A}{T\ulcorner A \urcorner}$$

In this case one can derive A as follows:

1	(1)	$T(\underline{k})$	Assumption
1	(2)	$T(\underline{k}) \rightarrow A$	1 and (*)
1	(3)	A	2, 3 (MP)
	(4)	$T(\underline{k}) \rightarrow A$	1, 3 (I→)
	(5)	$T(\underline{k})$	4 and (**)
	(6)	A	4, 5 (MP)

Notice that the result is not a counterexample to MV. In fact, according to Tarski's theorem about the inexpressibility of the truth predicate in a language that allows self-reference, the rules (*) and (**) cannot count as classical logical rules.

6 Some Concluding Remarks

In works subsequent to LP, Priest leaves the material conditional and tries to recover a "genuine" conditional. This is conceived of as a conditional that, by virtue of its logical meaning, the truth of the antecedent (independently of its possible falsity) implies the truth of the consequent (for a detailed discussion of such conditional, see [1, 2]).

We believe that the behaviour of the material conditional is, in some respects, more satisfactory than such genuine conditionals even for a dialetheist. In dealing with his genuine conditional, Priest tries to avoid Curry's paradox by means of a highly sophisticated modal semantics that invalidates the classical rule of contraction. This rule can be derived, in the system of natural deduction, from the basic rules of introduction and elimination of the conditional. So, by accepting MP, Priest is forced to reject the general validity of (I→). That seems a rash move: it strikes at the heart of hypothetical mathematical reasoning. Any working mathematician, when attempting to prove a conditional statement $A \rightarrow B$, assumes A and tries to prove B. If the attempt succeeds and you agree that her proof of B from A is correct but object that it does not count as a proof of $A \rightarrow B$, the most probable reply is that you do not understand the intended meaning of "if ...then".

Observe, in passing, that the rule (I→) is also fully valid for an intuitionist, who understands a proof of $A \rightarrow B$ as a method of transforming a proof of A into a proof of B. In cognitive terms, a proof of $A \rightarrow B$ must convince someone that, in order to recognize that B is true, it is sufficient to recognize that A is true. Nor does the rejection of (I→) seem to be dialetheically motivated by the possible presence of dialetheias. In fact, the understanding of $A \rightarrow B$ as expressing the condition that truth is preserved from A to B is by no means affected by the possibility that A or B could be a dialetheia. It seems, therefore, that accepting any proof of B from A as a proof of $A \rightarrow B$ is in accordance even with a dialetheic understanding of the conditional. It is of no use to object that some conditionals studied in certain branches of logic (strict, relevant or otherwise) do not validate (I→). Our point is that the usual understanding of the conditional occurring in mathematical sentences validates (I→) and is dialetheically intelligible. In addition, we have argued in Carrara et al. [1] that Priest's solution to Curry's paradox has nothing to do with dialetheism since no dialetheia plays any role in his solution. In contrast, our approach aims to recover a dialetheic interest for the material conditional. This provides a solution to Curry's paradox that exploits the presence of dialetheias and reduces it to the liar paradox.

References

1. Carrara, M., Martino, E., & Gaio, S. (2011). Can Priest's dialetheism avoid trivialism? In M. Pelis & V. Puncochar (Eds.), *The Logica Yearbook 2010* (pp. 53–64). London: College Publications.
2. Carrara, M., Martino, E., & Morato, V. (2012). On dialetheic entailment. In M. Pelis & V. Puncochar (Eds.), *The Logica Yearbook 2011* (pp. 37–48). London: College Publications.

3. Curry, H. B. (1942). The inconsistency of certain formal logics. *Journal of Symbolic Logic*, *7*, 115–7.
4. Priest, G. (1979). The logic of paradox. *Journal of Philosophical Logic*, *8*, 219–41.
5. Priest, G. (2006a). *Doubt truth to be a liar*. Oxford: Oxford University Press.
6. Priest, G. (2006b). *In contradiction*. Oxford: Oxford University Press.
7. Restall, G. (2013). Assertion, denial and non-classical theories. In F. Berto & E. Mares (Eds.), *Paraconsistency: Logic and applications* (pp. 81–99). Berlin: Springer.
8. Ripley, D. (2011). Negation, denial and rejection. *Philosophy Compass*, *6*(9), 622–9.

Strongly Semantic Information as Information About the Truth

Gustavo Cevolani

Abstract Some authors, most notably Luciano Floridi, have recently argued for a notion of "strongly" semantic information, according to which information "encapsulates" truth (the so-called "veridicality thesis"). We propose a simple framework to compare different formal explications of this concept and assess their relative merits. It turns out that the most adequate proposal is that based on the notion of "partial truth", which measures the amount of "information about the truth" conveyed by a given statement. We conclude with some critical remarks concerning the veridicality thesis in connection with the role played by truth and information as relevant cognitive goals of inquiry.

Keywords (Strongly) Semantic information · Truth · Misinformation · Veridicality thesis · Verisimilitude · Truthlikeness · Partial truth · Informative truth · Cognitive decision theory

1 Introduction

In recent years, philosophical interest in the concept of information and its logical analysis has been growing steadily (cf., e.g., [1, 2, 9, 13, 14, 22]). Philosophers and logicians have proposed various definitions of (semantic) information, and tried to elucidate the connections between this notion and related concepts like truth, probability, confirmation, and truthlikeness. While classical accounts, both in philosophy [4] and in (mathematical) information theory [32], define information in terms of (im)probability, more recent proposals try to link together information and truth (see, in particular, [10]). According to these proposals, the classical notion of

G. Cevolani (✉)
Department of Philosophy and Education, University of Turin, via S. Ottavio 20,
10124 Turin, Italy
e-mail: g.cevolani@gmail.com

R. Ciuni et al. (eds.), *Recent Trends in Philosophical Logic*,
Trends in Logic 41, DOI: 10.1007/978-3-319-06080-4_5,

information should be replaced, or at least supplemented with, a notion of "strongly semantic" information (henceforth, SSI), construed as well-formed, meaningful and "veridical" or "truthful" data about a given domain. This so-called "veridicality thesis" would imply "that 'true information' is simply redundant and 'false information', i.e., misinformation, is merely pseudo-information" [13, p. 82].[1] In this paper, we shall survey different formal explications of SSI, explore their conceptual relationships, and highlight their implications for the debate about the veridicality thesis triggered by Floridi's definition of SSI.

In Sect. 2, we review the "classical" definition of semantic information due to Carnap and Bar-Hillel [4] in the light of the critiques that it has received. The notion of SSI, and the related veridicality thesis, is discussed in Sect. 3. In Sect. 4, we survey three recent proposals, including Floridi's one, that define SSI in terms of different combinations of truth and information. We introduce a simple framework which allows us to compare these proposals, and argue in favor of one of them, which identifies SSI with the amount of information about "the truth" conveyed by a given statement. On this basis, in Sect. 5 we conclude that, in order to define sound notions of (true) information and misinformation, one can safely dispense with the veridicality thesis, that can be however accepted as a thesis concerning the epistemic goals guiding rational inquiry and cognitive decision making.

2 Information and Truth

The classical theory of (semantic) information [4, 32] is based on what Jon Barwise [1, p. 491] has called the "Inverse Relationship Principle" (IRP), i.e., the intuition that "eliminating possibilities corresponds to increasing information" [1, p. 488].[2] If A is a statement in a given language, IRP amounts to say that the information content of A can be represented as the set of (the linguistic descriptions of) all the possible state of affairs or "possible worlds" which are excluded by, or incompatible with, A. Accordingly, the amount of information conveyed by A will be proportional to the cardinality of that set. For the sake of simplicity, let us consider a finite propositional language $\mathcal{L}_n$ with n logically independent atomic sentences $p_1, \ldots, p_n$.[3] An atomic sentence p_i and its negation $\neg p_i$ are called "basic sentences" or "literals" of $\mathcal{L}_n$. Within $\mathcal{L}_n$, possible worlds are described by the so-called constituents of $\mathcal{L}_n$, which are conjunctions of n literals, one for each atomic sentence. Note that the set $\mathcal{C}$ of the constituents of $\mathcal{L}_n$ includes $q = 2^n$ elements and that only one of them, denoted by "$C_\star$", is true; thus, $C_\star$ can be construed as "the (whole) truth" in $\mathcal{L}_n$, i.e., as the complete true description of the actual world w.

---

[1] As a terminological remark, note that while "misinformation" simply denotes false or incorrect information, "disinformation" is false information deliberately intended to deceive or mislead.

[2] This idea can be traced back at least to Karl Popper [27, in particular Sects. 34 and 35, and Appendix IX, p. 411, footnote 8]; cf. also [3, p. 406].

[3] Nothing substantial, in what follows, depends on such assumption.

Given an arbitrary statement A of $\mathscr{L}_n$, let $R(A)$ be the the "range" of A, i.e., the set of constituents which entail A, corresponding to the set of possible worlds in which A is true (cf. [3, Sect. 18]). Then, the (semantic) information content of A is defined as:

$$Cont(A) \stackrel{\text{df}}{=} \mathscr{C} \setminus R(A) = R(\neg A). \tag{1}$$

A definition of the amount of information content of A can be given assuming that a probability distribution p is defined over the sentences of $\mathscr{L}_n$ [4, p. 15]:

$$cont(A) \stackrel{\text{df}}{=} 1 - p(A) = p(\neg A). \tag{2}$$

In agreement with IRP, the information conveyed by A is thus inversely related to the probability of A.[4]

Two immediate consequences of definitions in 1 and 2 are here worth noting. First, if $\top$ is an arbitrary logical truth of $\mathscr{L}_n$, then

$$Cont(\top) = \varnothing \text{ and } cont(\top) = 0 \tag{3}$$

since $\top$ is true in all possible worlds ($R(\top) = \mathscr{C}$) and hence $p(\top) = 1$. Second, if $\bot$ is an arbitrary logically false statement of $\mathscr{L}_n$, then

$$Cont(\bot) = \mathscr{C} \text{ and } cont(\bot) = 1 \tag{4}$$

since $\bot$ is false in all possible worlds ($R(\bot) = \varnothing$) and hence $p(\bot) = 0$. In short, tautologies are the least informative, and contradictions the most informative, statements of $\mathscr{L}_n$.

As D'Agostino and Floridi [7, p. 272] note, results 3 and 4 point to "two main difficulties" of the classical theory of semantic information as based on IRP. The first one is what Hintikka [20, p. 222] called "the scandal of deduction": since in classical deductive logic conclusion C is deducible from premises $P_1, \ldots, P_n$ if and only if the conditional $P_1 \wedge \cdots \wedge P_n \rightarrow C$ is a logical truth, to say that tautologies are completely uninformative is to say that logical inferences never yield an increase of information. This is another way of saying that deductive reasoning is "non-ampliative", i.e., that conclusion C conveys no information besides that contained in the premises P_i.[5] In this paper, we shall be concerned only with the second difficulty, called "the Bar-Hillel-Carnap semantic Paradox (BCP)" by Floridi [10, p. 198].

[4] In the literature, it is usual to say that Eqs. (1) and (2) define the (amount of) "substantive information" or "information content" of A, as opposed to the "unexpectedness" or "surprise value" of A, which is defined as $\mathrm{inf}(A) = -\log p(A)$ [4, p. 20]. On this distinction, see for instance Hintikka [18, p. 313] and Kuipers [22, p. 865].

[5] To hush up this "scandal", Hintikka [19] developed a distinction (for polyadic first-order languages) between "depth" and "surface" information, according to which logical truths may contain a positive amount of (surface) information (cf. also [31]).

Already Carnap and Bar-Hillel [4, pp. 7–8], while defending result 4 as a perfectly acceptable consequence of their theory, pointed out that it is *prima facie* counterintuitive:

> It might perhaps, at first, seems strange that a self-contradictory sentence, hence one which no ideal receiver would accept, is regarded as carrying with it the most inclusive information. It should, however, be emphasized that semantic information is here not meant as implying truth. A false sentence which happens to say much is thereby highly informative in our sense. Whether the information it carries is true or false, scientifically valuable or not, and so forth, does not concern us. A self-contradictory sentence asserts too much; it is too informative to be true.

As made clear in the above quote, BCP follows from the assumption that truth and information are independent concepts, in the sense that A doesn't need to be true in order to be informative—the so-called assumption of "alethic neutrality" (AN) [11, p. 359].[6]

Many philosophers (e.g. [9, pp. 41 ff]) have noted that AN is at variance with the ordinary use of the term "information", which is often employed as a synonym of "true information". In fact, we are used to say that "[a] person is 'well-informed' when he or she knows much—and thereby is aware of many truths" [24, p. 155]. On the other hand, AN appears more acceptable as far as other common uses of this term are concerned, for instance when we speak of the "information" processed by a computer [24]. Thus, linguistic intuitions are insufficient to clarify the question whether information and truth are or not independent. Some scholars, most notably Luciano Floridi [10–13], have forcefully argued that AN should be rejected in favor of the so-called "veridicality thesis" (VT), according to which genuine information has to be (at least approximately) true. According to VT, BCP would be solved since contradictions, being a paradigmatic case of false statements, are not informative at all. What exactly VT implies for the classical definition of semantic information is however unclear, and will be discussed in the next section.

3 What is "Strongly" Semantic Information?

Under its weakest reading, VT simply says that truth and information are not independent concepts, as the classical theory of semantic information assumes, and that an adequate theory of strongly semantic information (SSI) has to take both concepts into account. According to Floridi, VT says, more precisely, that "information encapsulates truth" [10, p. 198] in the sense that A has to be "truthful" [11, p. 366] or "veridical" [13, p. 105] in order to be informative at all. The underlying intuition is expressed by Dretske [9, p. 44–45] in these terms:

[6] This does not mean that this assumption is the only culprit. As noted during discussion at the *Trends in Logic XI* conference, a way of avoiding BCP would be to adopt a non-classical logic according to which contradictions do not entail everything and hence are not maximally informative. Systems of this kind are provided by those "connexive logics" that reject the classical principle *ex contradictione quodlibet* (for all A, $\bot$ entails A) in favor of the (Aristotelian) intuition that *ex contradictione nihil sequitur* (cf. [33, Sect. 1.3]).

> If everything I say to you is false, then I have given you no information. At least I have given you no information of the kind I purported to be giving. [...] In this sense of the term, *false* information and *mis*-information are not kinds of information—any more than decoy ducks and rubber ducks are kinds of ducks.

Taking this idea at face value, VT would imply that only true statements are informative. In turn, this would amount to define SSI as follows (cf. [12, p. 40]):

$$A \text{ is a piece of SSI iff } A \text{ is true.} \tag{5}$$

By definition, A is true iff $C_\star \in R(A)$. The (amount of) SSI conveyed by A would be still defined by $Cont(A)$ and $cont(A)$, but only true statements would be allowed to occur within (1) and (2).

Condition (5) is a straightforward formulation of the thesis that "information encapsulates truth", but it is doubtful that supporters of VT would be ready to subscribe to it. In fact, it implies that *all* false statements are plainly uninformative. As a consequence, (5) provides a solution of BCP, but too a strong one, which is at least as counterintuitive as BCP itself. In fact, both in science and in ordinary contexts any piece of information at disposal is arguably at best approximate and, strictly speaking, false. For example, if one says that "Rudolf Carnap was an influential philosopher of science born in Germany in 1890", while the correct date of birth is 1891, it seems strange to say that this false statement conveys the same amount of information as "Carnap was German or not German" and "Carnap was German and not German", i.e., no information at all. Examples of this kind seems sufficient to exclude (5) as a possible definition of SSI.

Sequoiah-Grayson [30] has argued that the crucial intuition underlying the notion of SSI is that A has to provide some "factual" or "contingent" information to count as a piece of information at all. This would amount to define SSI in terms of the following "contingency requirement" [30, p. 338]:

$$A \text{ is a piece of SSI iff } A \text{ is factual.} \tag{6}$$

Note that A is factual or contingent iff $\varnothing \neq R(A) \subset \mathscr{C}$. According to this view, only tautologies and contradictions are completely uninformative. Thus, BCP is solved by defining SSI not as true, but as factual information. It follows that both false and true contingent statements are informative after all. In particular, as Floridi [10, p. 206] notes[7]:

> two [statements] can both be false and yet significantly more or less distant from the event or state of affairs w about which they purport to be informative, e.g. "there are ten people in the library" and "there are fifty people in the library", when in fact there are nine people in the library. Likewise, two [statements] can both be true and yet deviate more or less significantly from w, e.g. "there is someone in the library" versus "there are 9 or 10 people in the library".

---

[7] In the following, we replace, without any significant loss of generality, Floridi's talk of "infons"—"discrete items of factual information qualifiable in principle as either true or false, irrespective of their semiotic code and physical implementation" [10, p. 199]—by talk of sentences or statements in the given language $\mathscr{L}_n$.

> This implies that a falsehood with a very low degree of discrepancy may be pragmatically preferable to a truth with a very high degree of discrepancy [28].

The above quotation highlights the strict link between the notion of SSI and that of verisimilitude or truthlikeness, construed, in Popperian terms, as similarity or closeness to "the whole truth" about a given domain (cf. also [2, pp. 90–91]). The idea of defining SSI in terms of verisimilitude has been indeed proposed (*ante litteram*) by Frické [15] and independently by D'Alfonso [8]; indeed, as Frické [15, p. 882] notes, this proposal explains how true and false statements can be both informative:

> With true statements, verisimilitude increases with specificity and comprehensiveness, so that a highly specific and comprehensive statement will have high verisimilitude; such statements also seem to be very informative. With false statements, verisimilitude is intended to capture what truth they contain; if false statements can convey information, and the view taken here is that they can, it might be about those aspects of reality to which they approximate. Verisimilitude and a concept of information appear to be co-extensive.

Thus, the amount of SSI that a (true or false) contingent statement A conveys will depend on how good an approximation A is to the actual world w (or to the true constituent $C_\star$). To make this idea precise, Floridi [10, Sect. 5, pp. 205–206 in particular] has proposed five conditions that an adequate notion of SSI should fulfil. Departing a little from Floridi's original formulation, they can be phrased as follows:

(SSI1) the true constituent $C_\star$ is maximally informative, since it is the complete true description of the actual world w

(SSI2) tautologies are minimally informative, since they do not convey factual information about w

(SSI3) contradictions are minimally informative, since they do not convey, so to speak, valuable information about w

(SSI4) false factual statements are more informative than contradictions

(SSI5) true factual statements are more informative than tautologies.

Note that SSI3 is required in order to avoid BCP. From SSI1 and SSI5 it follows that $C_\star$ is the most informative statement among all factual truths. Another requirement that can be defended as an adequacy condition for a notion of SSI is the following:

(SSI6) some false factual statements may be more informative than some true factual statements.

In fact, as made clear by Floridi's quotation above, a false statement may be a better approximation to the truth about w than a true one. Characterizing SSI by means of requirements SSI1–6 still leaves open the problem of how to define a rigorous counterpart of this notion, and in particular of how to quantify the amount of SSI conveyed by different statements. In the next section, we shall review and compare different measures of SSI proposed in the literature to address this issue.

4 A Basic Feature Approach to Strongly Semantic Information

Some authors have recently proposed different formal explications of the notion of SSI, in the form of measures of the degree or amount of SSI conveyed by statements of $\mathscr{L}_n$ [5, 8, 10, 13]. According to all these proposals, the degree of SSI of A is high, roughly, when A conveys much true information about w. A simple way of clarifying and comparing these measures is given by the so-called "basic feature" approach to verisimilitude [6] or "BF-approach" for short.

4.1 The Basic Feature Approach to Verisimilitude

According to the BF-approach, the "basic features" of the actual world w are described by the basic sentences or literals of $\mathscr{L}_n$. A conjunctive statement, or c-statement for short, is a consistent conjunction of k literals of $\mathscr{L}_n$, with $k \leq n$.[8] The "basic content" of a c-statement A is the set $b(A)$ of the conjuncts of A: each member of this set will be called a "(basic) claim" of A. One can check that $\mathscr{L}_n$ has exactly 3^n c-statements, including the "tautological" c-statement with $k = 0$ and the 2^n constituents with $k = n$. Indeed, note that $C_\star$ itself is a c-statement, being the conjunction of the true basic sentences in $\mathscr{L}_n$, i.e., the most complete true description of the basic features of w.

When A is compared to $C_\star$, $b(A)$ is partitioned into two subsets: the set $t(A, C_\star)$ of the true claims of A and the set $f(A, C_\star)$ of the false claims of A. Let us call each element of $t(A, C_\star)$ a *match*, and each element of $f(A, C_\star)$ a *mistake* of A. Note that A is true when $f(A, C_\star) = \varnothing$, i.e., when A doesn't make mistakes, and false otherwise. Moreover, A is "completely false" when $t(A, C_\star) = \varnothing$, i.e., when A makes *only* mistakes. For the sake of notational simplicity, let us introduce the symbols k_A, t_A, and f_A to denote, respectively, the number of claims, of matches, and of mistakes, of A—i.e., the cardinalities of $b(A)$, $t(A, C_\star)$, and $f(A, C_\star)$, respectively. The *degree* of basic content $cont_b(A)$, of true basic content $cont_t(A, C)$, and of false basic content $cont_f(A, C)$, of A is defined as follows:

$$cont_b(A) \overset{\text{df}}{=} \frac{k_A}{n} \quad \text{and} \quad cont_t(A, C_\star) \overset{\text{df}}{=} \frac{t_A}{n} \quad \text{and} \quad cont_f(A, C_\star) \overset{\text{df}}{=} \frac{f_A}{n} \tag{7}$$

i.e., as the normalized number of claims, of matches, and of mistakes, made by A.

The number of matches of A divided by the total number of its claims represents an adequate measure for the (degree of) "accuracy" $acc(A)$ of a c-statement A:

$$acc(A) \overset{\text{df}}{=} \frac{t_A}{k_A} = \frac{cont_t(A, C_\star)}{cont_b(A)}. \tag{8}$$

---

[8] In logical parlance, a c-statement is a statement in conjunctive normal form such that each of its clauses is a single literal. Following Oddie [26, p. 86], a c-statement may be also called a "quasi-constituent", since it can be conceived as a "fragment" of a constituent.

Conversely, the (degree of) "inaccuracy" of A can be defined as

$$inacc(A) \stackrel{\text{df}}{=} \frac{f_A}{k_A} = \frac{cont_f(A, C_\star)}{cont_b(A)}. \tag{9}$$

As one can check, $inacc(A) = 1 - acc(A)$. Moreover, if A is true, then $t_A = k_A$ and $acc(A)$ receives its maximum value, i.e., 1; conversely, $inacc(A)$ is 0. When A is completely false, $acc(A) = 0$ and $inacc(A) = 1$. In sum, all true c-statements are maximally accurate, while all completely false c-statements are maximally inaccurate.

As Popper notes, the notion of verisimilitude "represents the idea of *approaching comprehensive truth.* It thus combines truth and content" [28, p. 237, emphasis added].[9] Thus, accuracy is only one "ingredient" of verisimilitude, the other being (information) content. In other words, we may say that a c-statement A is highly verisimilar if it says many things about the target domain, and if many of these things are true; in short, if A makes many matches and few mistakes about w. This intuition is captured by the following "contrast measure" of the verisimilitude of c-statements A [6, p. 188]:

$$vs_\phi(A) \stackrel{\text{df}}{=} cont_t(A, C_\star) - \phi cont_f(A, C_\star) \tag{10}$$

where $\phi > 0$.[10] Intuitively, different values of ϕ reflect the relative weight assigned to truths and falsehoods, i.e., to the matches and mistakes of A. Some interesting feature of this definition are the following. First, while all true A are equally accurate, since $acc(A) = 1$, they may well vary in their relative degree of verisimilitude. More precisely, vs_ϕ satisfies the Popperian requirement that verisimilitude co-varies with logical strength among truths[11]

$$\text{If } A \text{ and } B \text{ are true and } A \text{ entails } B, \text{ then } vs_\phi(A) \geq vs_\phi(B). \tag{11}$$

Thus, logically stronger truths are more verisimilar than weaker ones. This condition, however, doesn't hold amongst false statements, since logically stronger falsehoods may well lead us farther from the truth. In particular[12]

---

[9] For different accounts of verisimilitude, see [21, 24, 26, 29].

[10] One may note that measure vs_ϕ is not normalized, and varies between $-\phi$ and 1. A normalized measure of the verisimilitude of A is $(vs_\phi(A) + \phi)/(1 + \phi)$, which varies between 0 and 1.

[11] *Proof* note that, among c-statements, A entails B iff $b(A) \supseteq b(B)$. If both are true, this implies $t(A, C_\star) \supseteq t(B, C_\star)$ and hence $vs_\phi(A) = cont_t(A, C_\star) \geq cont_t(B, C_\star) = vs_\phi(B)$. For discussion of this Popperian requirement, see [24, pp. 186–187, 233, 235–236]. Note also that vs_ϕ satisfies the stronger requirement that among true theories, the one with the greater degree of (true) basic content is more verisimilar than the other; i.e., if A and B are true and $cont_t(A, C) > cont_t(B, C)$ then $vs_\phi(A) > vs_\phi(B)$. □

[12] *Proof* if A entails B and both are completely false, then $f(A, C_\star) \supseteq f(B, C_\star)$ and hence $cont_f(A, C_\star) \geq cont_f(B, C_\star)$. Since ϕ is positive, it follows that $vs_\phi(A) = -\phi cont_f(A, C_\star) \leq -\phi cont_f(B, C_\star) = vs_\phi(B)$. □

$$\text{If } A \text{ and } B \text{ are completely false and } A \text{ entails } B, \text{ then } vs_{\phi}(A) \leq vs_{\phi}(B). \tag{12}$$

In words, logically *weaker* complete falsehoods are better than stronger ones. Finally, note that a false c-statement may well be more verisimilar than a true one; however, completely false c-statements are always less verisimilar than true ones.[13]

4.2 Quantifying Strongly Semantic Information

Measures of true (basic) content, of (in)accuracy, and of verisimilitude, or combinations of them, have all been proposed as formal explicata of the notion of SSI. For instance, according to Floridi [10], SSI may be construed as a combination of content and accuracy. More precisely, among truths, SSI increases with content, or, better, it decreases with the degree of "vacuity" of A, defined as the normalized cardinality of the range of A:

$$vac(A) = \frac{|R(A)|}{|\mathscr{C}|}. \tag{13}$$

Among falsehoods, SSI increases with accuracy, and decreases with inaccuracy. Moreover, $C_\star$ is assigned the highest degree of SSI, and contradictions the lowest. In sum, Floridi's measure of SSI is defined as follows [10, pp. 208–210]:

$$cont_S(A) \stackrel{\text{df}}{=} \begin{cases} 1 - vac(A)^2 & \text{if } A \text{ is true and } A \not\equiv C_\star \\ 1 - inacc(A)^2 & \text{if } A \text{ is factually false} \\ 1 & \text{if } A \equiv C_\star \\ 0 & \text{if } A \text{ is contradictory.} \end{cases} \tag{14}$$

One can easily check that Floridi's measure satisfies all requirements SSI1–6 (SSI1 and SSI3 are fulfilled by stipulation). In particular, since tautologies have maximal degree of vacuity, their degree of SSI is 0 (cf. SSI2). Moreover, $cont_S$ satisfies the Popperian requirement (11): if A and B are true and A entails B, then $cont_S(A) \geq cont_S(B)$.

A simpler formulation of (14) is given by the following measure:

$$cont^*_S(A) \stackrel{\text{df}}{=} \begin{cases} cont_b(A) = cont_t(A, C_\star) & \text{if } A \text{ is true} \\ acc(A) & \text{if } A \text{ is false.} \end{cases} \tag{15}$$

[13] *Proof* If A is true, then $vs_\phi(A) = cont_t(A, C_\star)$; if B is completely false, then $vs_\phi(B) = -\phi cont_f(B, C_\star)$; since ϕ is positive, it follows that $vs_\phi(A) > vs_\phi(B)$. □

One can check that $cont^*_S$ and $cont_S$ are ordinally equivalent in the sense that, given any two c-statements A and B, $cont^*_S(A) \gtreqless cont^*_S(B)$ iff $cont_S(A) \gtreqless cont_S(B)$.[14] Thus, also the $cont^*_S$ measure satisfies requirements SSI1–6.[15]

According to (15), $cont^*_S$ increases with the degree of (true) basic content among truths, and with accuracy among falsehoods. In this connection, it may be worth noting that $acc(A)$ is a straightforward measure of the "approximate truth" of A, construed as the closeness of *A to being true* [24, pp. 177, 218]. In fact, if A is true, then $acc(A) = 1$; while if A is false, then $acc(A)$ is smaller than 1 and increases the closer A is to being true. Recalling that, if A is completely false, $inacc(A) = 1$, *inacc* can be construed as a measure of the closeness of *A to being completely false*. Since all completely false c-statements are equally (and maximally) inaccurate, if A is completely false then $cont_S(A) = 0$, i.e., A conveys no SSI about w.

Following the idea that SSI is a combination of content and accuracy, D'Alfonso [8] has proposed to use measures of verisimilitude for quantifying SSI. An immediate advantage is that a unique measure vs_ϕ is used to assess the degree of SSI of both true and false statements. One can check that vs_ϕ satisfies requirements SSI1—$vs_\phi(C_\star)$ is maximal, since "nothing is as close as the truth as the whole truth itself" [26, p. 11]—, and SSI4–6. Moreover, since vs_ϕ is undefined for contradictions, one can just stipulate that their degree of SSI is 0, in agreement with SSI3 (but see [8, p. 77]). However, as D'Alfonso acknowledges [8, p. 73], vs_ϕ violates SSI2, since tautologies are more verisimilar than some false statements. In particular, all completely false statements are less verisimilar than tautologies: in fact, when conceived as an answer to a cognitive problem, a tautology corresponds to suspending the judgment, which is better than accepting "serious" falsehoods. Another problem with vs_ϕ as a measure of SSI is that, among completely false c-statements, vs_ϕ decreases with content. This is perfectly natural as far as verisimilitude is concerned, but it seems at variance with the idea that "SSI encapsulates truth". In fact, a completely false c-statement is not "veridical" at all, in the sense that it conveys no true factual information about the world; accordingly, its degree of SSI should be 0.

In order to overcome these difficulties, Cevolani [5] suggested $cont_t$, the degree of true basic content, as a measure of SSI. Note that this amounts to ignore, in (15), the second half of Floridi's (rephrased) measure $cont^*_S$ and to use $cont_t$ as a measure of SSI for *both* true and false statements. The latter measure was proposed by Hilpinen [17] as an explication of the notion of "partial truth", measuring the amount of *information about the truth* conveyed by a (true or false) statement A (see also [24, Sects. 5.4 and 6.1]). Since all requirements SSI1–6 are satisfied by $cont_t$,

---

[14] *Proof sketch* When A is a c-statement, the constituents in its range are 2^{n-k_A}; it follows that $vac(A) = 2^{n-k_A}/2n$, i.e., $1/2^{k_A}$. Thus, among true c-statements, $cont_S(A) = 1 - 1/2^{k_A}$ co-varies with the degree of basic content of A, $b(A) = k_A/n$. As far as false c-statements are concerned, since $inacc(A) = 1 - acc(A)$, $cont_S$ co-varies with the accuracy of A. □

[15] Note that, by definition, a c-statement can not be contradictory; hence, $cont^*_S$ is undefined for contradictions. Of course, it is always possible to stipulate, as Floridi does, that contradictions have a minimum degree of SSI.

this is in fact an adequate measure of SSI.[16] In particular, it delivers a minimum degree of SSI for both tautologies and completely false statements, in agreement with Floridi's measure $cont_S$. Moreover, $cont_t$ increases with content among truths, whereas for false statements it depends on how much true information they convey (cf. [24, p. 176]).

While $cont_S$ and $cont_t$ are ordinally equivalent as far as true statements are concerned, they differ in assessing the degree of SSI conveyed by false statements. If A is a false c-statement, $cont_S(A)$ measures the accuracy or closeness to be true of A, i.e., increases with t_A/k_A, whereas $cont_t(A, C_\star)$ increases with the amount of information about the truth conveyed by A, i.e., increases with t_A/n. As an example, assume that $C_\star \equiv p_1 \wedge \cdots \wedge p_n$ and let A and B be, respectively, the c-statements $p_1 \wedge \neg p_2$ and $p_1 \wedge p_2 \wedge \neg p_3 \wedge \neg p_4$. Then, $cont_S(A) = cont_S(B) = \frac{1}{2}$: according to Floridi, the degree of SSI of A and B is the same, since they are equally accurate. In this sense, $cont_S$ is, so to speak, insensitive to content as far as false statements are concerned (cf. [24, p. 219]). On the other hand, $cont_t(A, C_\star) = \frac{1}{n}$ is smaller than $cont_t(B, C_\star) = \frac{2}{n}$, since B makes two matches instead of one, i.e., conveys more information about the truth than A. Thus, $cont_t$ appears as a more adequate *information* measure than $cont_S$.

In this connection, one may note that $cont_t$ is insensitive to the number of mistakes contained in a false c-statement.[17] For instance, assume again that $C_\star \equiv p_1 \wedge \cdots \wedge p_n$ is the truth and that A is the false c-statement $p_1 \wedge \neg p_2$. If B is obtained from A by adding to it a false claim, for instance if $B \equiv p_1 \wedge \neg p_2 \wedge \neg p_3$, its degree of partial truth does not change, since $cont_t(A, C_\star) = cont_t(B, C_\star) = \frac{1}{n}$. However, B is now less accurate than A: accordingly, $cont_S(B) = \frac{1}{3} < \frac{1}{2} = cont_S(A)$. While it may appear counterintuitive that $cont_t(A, C_\star)$ does not decrease when the number of mistakes made by A increases, this is just another way of saying that $cont_t$ measures informativeness about the truth and not accuracy (nor verisimilitude). In other words, in the example above it is only relevant that both A and B make one match, independently from the number of their mistakes. At a deeper level, this depends on the following feature of $cont_t$:

$$\text{If } A \text{ entails } B, \text{ then } cont_t(A, C_\star) \geq cont_t(B, C_\star). \tag{16}$$

This result says that strengthening a c-statement (i.e., adding to it true or false new claims) never yields a decrease of its degree of partial truth (cf. [24, p. 220]). In turn, comparing result (16) with the Popperian requirement (11) explains the difference between a measure of information about the truth, like $cont_t$, and a measure of closeness to the whole truth like vs_ϕ (for which, of course, (16) does not hold).

To sum up, we considered three ways of quantifying SSI: Floridi's $cont_S$ measure, the verisimilitude measure vs_ϕ, and the partial truth measure $cont_t$. We argued that

---

[16] Note again that $cont_t$ is undefined for contradictions, which can be assigned a minimum degree of SSI by stipulation. Interestingly, an argument to this effect was already proposed by Hilpinen [17, p. 30].

[17] I thank Gerhard Schurz for raising this point in discussion during the *Trends in Logic XI* conference.

both $cont_S$ and vs_ϕ are inadequate, for different reasons, in evaluating the SSI of false c-statements. Indeed, $cont_S$ cannot discriminate among differently informative but equally (in)accurate c-statements, while vs_ϕ favors less informative completely false c-statements over more informative ones, despite their being on a par relative to their true basic content (none). Thus, we submit that the notion of partial truth, as defined by the $cont_t$ measure, provides the most adequate explication of the concept of SSI, which should be conceived as the amount of information about the truth conveyed by a given statement.

4.3 Generalizing the Basic Feature Approach

One may complain that the approach presented in this section is unsatisfactory since it is restricted to a very special kind of sentences in a formal language, i.e., "conjunctive" statements. Indeed, all the notions considered in the previous discussion, including the $cont_t$ measure, are undefined for statements which can not be expressed as conjunctions of literals. However, our approach can be easily generalized to any language characterized by a suitable notion of constituent—or state description in the sense of Carnap [3]—including first-order monadic and polyadic languages [24] and second-order languages [26]. In such "languages with constituents", any non-contradictory sentence A can be expressed as the disjunction of the constituents entailing A (describing the possible worlds where A is true), i.e., in its so-called normal disjunctive form:

$$A \equiv \bigvee_{C_i \in R(A)} C_i. \tag{17}$$

If one assumes that a "distance function" $\Delta(C_i, C_j)$ is defined between any two constituents C_i and C_j,[18] one can (re)define the notion of partial truth for arbitrary statements as follows [24, pp. 217 ff.]:

$$pt(A) \stackrel{\text{df}}{=} 1 - \Delta_{max}(A, C_\star) = 1 - \max_{C_i \in R(A)} \Delta(C_i, C_\star). \tag{18}$$

Intuitively, $pt(A)$ is high when A excludes possible worlds which are far from the truth. Note that, if A is a c-statement, $pt(A) = cont_t(A, C_\star)$: i.e., partial truth as defined above generalizes the notion of degree of true basic content as defined in (7).[19]

[18] Usually, $\Delta(C_i, C_j)$ is identified with the so-called normalized Hamming distance (or Dalal distance, as it is also known in the field of AI), i.e., with the number of literals on which C_i and C_j *disagree*, divided by the total number n of atomic sentences.

[19] *Proof* Note that, if A is a c-statement, all constituents C_i in the range of A (which are c-statements themselves) are "completions" of A in the sense that $b(A) \subset b(C_i)$. The constituent in $R(A)$ farthest from $C_\star$ will be the one which makes all possible additional mistakes besides the mistakes already made by A: this means that $\Delta_{max}(A, C_\star) = 1 - \frac{t_A}{n}$. It follows by (7) that $pt(A) = 1 - (1 - \frac{t_A}{n}) = cont_t(A, C_\star)$. □

5 Conclusions: Information and Truth Revisited

Should only true contingent statements count as pieces of information after all? The recent debate on the veridicality thesis, triggered by Floridi's definition of SSI, has not reached a consensus on this point (see [14, Sect.3.2.3], for a survey of the main contributions).

The supporters of VT argue "that 'true information' is simply redundant and 'false information', i.e. misinformation, is merely pseudo-information" [11, p. 352]. According to this view, to come back to our example in Sect. 3, the statement "Rudolf Carnap was an influential philosopher of science born in Germany in 1890" would not be a piece of information at all, but may perhaps be split in two parts: an informative one—i.e.,"Rudolf Carnap was an influential philosopher of science born in Germany"—and a non-informative one—i.e., "Rudolf Carnap was born in 1890" (cf. [11, p. 361]). In short, this would amount to introduce a distinction between "information" and (say) "semantic content", the latter being the alethically neutral concept analized by Carnap and Bar-Hillel [4]. Accordingly, "information" would denote true semantic content and "misinformation" false semantic content (with tautologies and contradictions construed as extreme special cases of this twofold classification). Such a strategy would be perhaps in line with some analyses in the pragmatics of natural languages, like Paul Grice's study of "conversational implicatures" (cf. [11, p. 366]). In particular, the so-called "maxim of quality" of effective communication—"Do not say what you believe to be false" [16, p. 26]—apparently implies that "false information is not an inferior kind of information; it just is not information" [16, p. 371].

In this paper, we followed the opposite strategy of rejecting VT, and treat information and truth as independent notions, in agreement with the classical view of Carnap and Bar-Hillel [4]. The guiding idea of our discussion has been that SSI should not be construed as a new, more adequate explication of the notion of semantic information, but expresses the amount of (classical) semantic information *about the truth* conveyed by a given statement. According to this view, contingent statements are always informative but may be more or less successful in conveying true information about the world. This becomes especially clear as far as c-statements are concerned. In fact, (true and false) c-statements are the more informative about the truth the more true claims, or matches, they make about the world. In particular, completely false c-statements do not convey any information about the truth, and, in this sense, are plainly uninformative since they are not even "partially" true. This is a way of making sense of Dretske's remark that "If *everything* I say to you is false, then I have given you no information" [9, p. 44, emphasis added]. Finally, tautologies are also uninformative about the truth, since they do not convey any amount of factual information.

In this connection, one should note that the present approach also provides a straightforward quantitative definition of misinformation, a task that Floridi [10, p. 217] left for subsequent research. In fact, an adequate measure $misinf(A)$ of the misinformation conveyed by a c-statement A is given by its degree of false basic content defined in (7):

$$misinf(A) \stackrel{\text{df}}{=} cont_f(A, C_\star) = \frac{f_A}{n} \tag{19}$$

i.e., by the normalized number of the mistakes made by A. Since $cont_t(A, C_\star) + cont_f(A, C_\star) = cont_b(A)$, misinformation and partial truth (or information about the truth) are, so to speak, "complementary" notions.[20]

Finally, the fact that truth and information are here treated as independent concepts should not obscure an important point. As emphasized by philosophers of science and cognitive decision theorists—like Popper [28], Levi [23], Hintikka [18, 20], Kuipers [21], and Niiniluoto [24, 25], among others—both truth *and* information are important goals of rational (scientific) inquiry [25, Sect. 3.4]. In other words, at least according to any minimally realist view of science and ordinary knowledge, among the "epistemic utilities" guiding inquiry both truth and information have to play a prominent role. As Niiniluoto [25] notes, if "truth and nothing but the truth" were the only relevant aim of inquiry, then one should accept, as the best hypotheses at disposal, those statements which are more likely to be true—i.e., more probable—given the available evidence. This recommendation would lead to the "extremely conservative policy" of preferring tautologies, as well as statements logically implied by the evidence, over any other available hypothesis. On the other hand, if information were the only relevant epistemic utility, then, due to BCP (cf. Eq. 4), one should always accept contradictory hypotheses. Thus, as Levi [23] made clear, an adequate account of the cognitive goals of inquiry requires some notion of *informative truth*—i.e., some combination of both the truth value and the information content of alternative hypotheses.

The different measures considered in Sect. 4 can all be construed as different explications of this notion of informative truth. Accordingly, SSI can be conceived as a particular kind of epistemic utility combining truth and information, i.e., partial truth. In turn, one can interpret VT as a thesis concerning not information itself, but a corresponding appropriate notion of epistemic utility. In other words, while VT can be safely rejected as far as the *definition* of semantic information is concerned, the idea that "information encapsulates truth" can be accepted as a thesis about the ultimate cognitive goals guiding rational inquiry.

Acknowledgments This paper is based on presentations delivered at the *Fourth Workshop on the Philosophy of Information* (University of Hertfordshire, 10–11 May 2012) and at the *Trends in Logic XI* conference (Ruhr University Bochum, 3–5 June 2012). I thank the participants in those meetings, and in particular Luciano Floridi and Gerhard Schurz, for valuable feedback. This work was supported by Grant CR 409/1-1 from the Deutsche Forschungsgemeinschaft (DFG) as part of the priority program *New Frameworks of Rationality* (SPP 1516) and by the Italian Ministry of

[20] This is still clearer if one consider the generalization of the definition above to arbitrary (non-conjunctive) statements A. Given (18), the misinformation conveyed by A is given by $1 - pt(A) = \Delta_{max}(A, C_\star)$, that reduces to $misinf(A)$ as far as c-statements are concerned (the proof is straightforward, see footnote 19).

Scientific Research within the FIRB project Structures and dynamics of knowledge and cognition (Turin unit: D11J12000470001).

References

1. Barwise, J. (1997). Information and impossibilities. *Notre Dame Journal of Formal Logic*, *38*(4), 488–515.
2. Bremer, M., & Cohnitz, D. (2004). *Information and information flow: An introduction*. Frankfurt: Ontos Verlag.
3. Carnap, R. (1950). *Logical foundations of probability*. Chicago: University of Chicago Press.
4. Carnap, R., & Bar-Hillel, Y. (1952). *An outline of a theory of semantic information*. Technical Report 247, MIT Research Laboratory of Electronics.
5. Cevolani, G. (2011). Strongly semantic information and verisimilitude. *Ethics & Politics*, *2*, 159–179. http://www2.units.it/etica/
6. Cevolani, G., Crupi, V., & Festa, R. (2011). Verisimilitude and belief change for conjunctive theories. *Erkenntnis*, *75*(2), 183–202.
7. D'Agostino, M., & Floridi, L. (2009). The enduring scandal of deduction. *Synthese*, *167*(2), 271–315.
8. D'Alfonso, S. (2011). On quantifying semantic information. *Information*, *2*(1), 61–101.
9. Dretske, F. (1981). *Knowledge and the flow of information*. Cambridge: MIT Press.
10. Floridi, L. (2004). Outline of a theory of strongly semantic information. *Minds and Machines*, *14*(2), 197–221.
11. Floridi, L. (2005). Is semantic information meaningful data? *Philosophy and Phenomenological Research*, *70*(2), 351–70.
12. Floridi, L. (2007). In defence of the veridical nature of semantic information. *European Journal of Analytic Philosophy*, *3*(1), 31–1.
13. Floridi, L. (2011a). *The philosophy of information*. Oxford: Oxford University Press.
14. Floridi, L. (2011b). Semantic conceptions of information. In E. N. Zalta (Ed.), *The Stanford encyclopedia of philosophy* (Spring 2011 ed.). http://plato.stanford.edu/archives/spr2011/entries/information-semantic/
15. Frické, M. (1997). Information using likeness measures. *Journal of the American Society for Information Science*, *48*(10), 882–892.
16. Grice, H. P. (1989). *Studies in the way of words*. Cambridge: Harvard University Press.
17. Hilpinen, R. (1976). Approximate truth and truthlikeness. In M. Przełecki, K. Szaniawski & R. Wójcicki (Eds.), *Formal methods in the methodology of the empirical sciences* (pp. 19–42). Dordrecht: Reidel.
18. Hintikka, J. (1968). The varieties of information and scientific explanation. In B. V. Rootselaar & J. Staal (Eds.), *Logic, methodology and philosophy of science* III (Vol. 52, pp. 311–331). Amsterdam: Elsevier.
19. Hintikka, J. (1970). Surface information and depth information. In J. Hintikka & P. Suppes (Eds.), *Information and inference*, (pp. 263–297). Dordrecht: Reidel.
20. Hintikka, J. (1973). *Logic, language-games and information*. Oxford: Oxford University Press.
21. Kuipers, T. A. F. (2000). *From instrumentalism to constructive realism*. Dordrecht: Kluwer Academic Publishers.
22. Kuipers, T. A. F. (2006). Inductive aspects of confirmation, information and content. In R. E. Auxier & L. E. Hahn (Eds.), *The philosophy of Jaakko Hintikka* (pp. 855–883). Chicago and La Salle: Open Courts.
23. Levi, I. (1967). *Gambling with truth*. New York: Alfred A. Knopf.
24. Niiniluoto, I. (1987). *Truthlikeness*. Dordrecht: Reidel.
25. Niiniluoto, I. (2011). Scientific progress. In E. N. Zalta (Ed.), *The Stanford encyclopedia of philosophy* (Summer 2011 ed.). http://plato.stanford.edu/entries/scientific-progress/

26. Oddie, G. (1986). *Likeness to truth*. Dordrecht: Reidel.
27. Popper, K. R. (1934). *Logik der Forschung*. Vienna: Julius Springer [revised edition: *The logic of scientific discovery*. London: Routledge, 2002 (Hutchinson, London, 1959)].
28. Popper, K. R. (1963). *Conjectures and refutations* (3rd ed.). London: Routledge and Kegan Paul.
29. Schurz, G., & Weingartner, P. (2010). Zwart and franssen's impossibility theorem holds for possible-world-accounts but not for consequence-accounts to verisimilitude. *Synthese*, *172*, 415–436.
30. Sequoiah-Grayson, S. (2007). The metaphilosophy of information. *Minds and Machines*, *17*(3), 331–44.
31. Sequoiah-Grayson, S. (2008). The scandal of deduction. *Journal of Philosophical Logic*, *37*(1), 67–94.
32. Shannon, C. (1948). A mathematical theory of communication. *Bell System Technical Journal*, *27*, 379–423, 623–656.
33. Wansing, H. (2010). Connexive logic. In E. N. Zalta (Ed.), *The stanford encyclopedia of philosophy* (Fall 2010 ed.). http://plato.stanford.edu/entries/logic-connexive/

Priest's Motorbike and Tolerant Identity

Pablo Cobreros, Paul Egré, David Ripley and Robert van Rooij

Abstract In his chapter 'Non-transitive identity' [8], Graham Priest develops a notion of non-transitive identity based on a second-order version of LP. Though we are sympathetic to Priest's general approach to identity we think that the account can be refined in different ways. Here we present two such ways and discuss their appropriateness for a metaphysical reading of indefiniteness in connection to Evans' argument.

Keywords Logical consequence · Identity · Indeterminacy · Logic of paradox

P. Cobreros (✉)
Facultad de Filosofía y Letras, Universidad de Navarra, 31009 Pamplona, Spain
e-mail: pcobreros@unav.es

P. Egré (✉)
Institut Jean-Nicod, Département d'Etudes Cognitives, Ecole Normale Supérieure, 29, rue d'Ulm, Pavillon Jardin - 1er étage, 75005 Paris, France
e-mail: paulegre@gmail.com

D. Ripley (✉)
Department of Philosophy, University of Connecticut, 344 Mansfield Rd Storrs, Storrs, CT 06269, USA
e-mail: davewripley@gmail.com

R. van Rooij (✉)
Feculteit der Geesteswetenschappen, Institute for Logic, Language and Computation, University of Amsterdam, Oude Turfmarkt 143, 1012 GC Amsterdam, Netherland
e-mail: R.A.M.vanRooij@uva.nl

R. Ciuni et al. (eds.), *Recent Trends in Philosophical Logic*,
Trends in Logic 41, DOI: 10.1007/978-3-319-06080-4_6,

1 Priest's Motorbike and LP-Identity

1.1 Priest's Motorbike

Priest motivates his account of identity based on the following case:

> Suppose I change the exhaust pipes on my bike; is it or is it not the same bike as before? It is, as the traffic registration department and the insurance company will testify; but it is not, since it is manifestly different in appearance, sound, and acceleration.
>
> Dialecticians, such as Hegel, have delighted in such considerations, since they appear to show that the bike both is and is not the same. A standard reply here is to distinguish between the bike itself and its properties. After the change of exhaust pipes the bike is numerically the same bike; it is just that some of its properties are different. Perhaps, for the case at hand, this is the right thing to say. But the categorical distinction between the thing itself and its properties is one which is difficult to sustain; to suppose that the bike is something over and above all of its properties is simply to make it a mysterious *Ding an sich*. Thus, suppose that I change, not just the exhaust pipes, but, in succeeding weeks, the handle bars, wheels, engine, and in fact all the parts, until nothing of the original is left. It is now a numerically different bike, as even the traffic office and the insurance company will concur. At some stage, it has changed into a different bike, i.e. it has become a different machine: the bike itself is numerically different. [8, 406]

Other cases of this sort seem to show that identity fails to be transitive. There is an implicit link in the literature between the ideas that identity is transitive and that indeterminacy associated to vagueness is purely semantic. As David Lewis puts it:

> The reason it's vague where the outback begins is not that there's this thing, the outback, with imprecise borders; rather there are many things, with different borders, and nobody has been fool enough to try to enforce a choice of one of them as the official referent of the word 'outback'. Vagueness is semantic indecision [6, 213].

In the following section we review Priest's strategy to define a non-transitive notion of identity based on LP.

1.2 Second-Order LP and Identity

LP is a paraconsistent logic with a natural dialetheist interpretation: for some property P and thing a, a is both P and not-P. We can formulate LP's semantics in a very straightforward way making use of three values and a Strong-Kleene valuation schema (our presentation is different in style, but equivalent to Priest's [8]). More specifically, for a first-order language (just unary predicates and no complex terms) $\mathscr{L}$:

Definition 1 An *MV-model* is a structure $\langle D, \mathbb{I} \rangle$ such that:

- D a non-empty domain of quantification.
- $\mathbb{I}$ is an interpretation function:

- For a name or variable a, $\mathbb{I}(a) \in D$
- For any predicate P, $\mathbb{I}(P) \in \{1, \frac{1}{2}, 0\}^D$
- For an atomic formula Pa, $\mathbb{I}(Pa) = \mathbb{I}(P)\mathbb{I}(a)$
- $\mathbb{I}(\neg A) = 1 - \mathbb{I}(A)$
- $\mathbb{I}(A \wedge B) = min(\mathbb{I}(A), \mathbb{I}(B))$
- $\mathbb{I}(A \vee B) = max(\mathbb{I}(A), \mathbb{I}(B))$
- $\mathbb{I}(\exists x A) = max(\{\mathbb{I}'(A) : \mathbb{I}' \text{ is an } x\text{-variant of } \mathbb{I}\})$
- $\mathbb{I}(\forall x A) = min(\{\mathbb{I}'(A) : \mathbb{I}' \text{ is an } x\text{-variant of } \mathbb{I}\})$

Definition 2 We say that $\Gamma \models^{LP} \Delta$ iff there is no MV-model M such that $\mathbb{I}(A) > 0$, for every $A \in \Gamma$ and $\mathbb{I}(B) = 0$ for every $B \in \Delta$.

The material conditional $(A \supset B)$ is defined as $(\neg A \vee B)$ and the material biconditional $(A \equiv B)$ as $(A \supset B) \wedge (B \supset A)$.

Consider now the expansion of $\mathscr{L}$ to a language $\mathscr{L}_2$ including second-order variables and quantifiers. Our semantics should now take care of these, including a domain of possible values of second-order variables. More specifically:

Definition 3 An *MV_2-model* is a structure $\langle D_1, D_2, I\rangle$ such that:

- D_1 is a non-empty domain of quantification.
- D_2 is a set of functions in $\{1, \frac{1}{2}, 0\}^{D_1}$
- $\mathbb{I}$ is an interpretation function identical to that of MV-models except for second-order quantified statements:
  - $\mathbb{I}(\exists X A) = max(\{\mathbb{I}'(A) : \mathbb{I}' \text{ is an } X\text{-variant of } \mathbb{I}\})$
  - $\mathbb{I}(\forall X A) = min(\{\mathbb{I}'(A) : \mathbb{I}' \text{ is an } X\text{-variant of } \mathbb{I}\})$

We might want to impose certain constraints on D_2, like that for each $A \subseteq D_1$ there is an $f \in D_2$ such that $f(a) > 0$ for each $a \in A$. However, we won't force D_2 to contain all functions from D_1 to $\{1, \frac{1}{2}, 2\}$ [8, 408].

Definition 4 We say that $\Gamma \models_2^{LP} \Delta$ iff there is no MV_2-model M such that $\mathbb{I}(A) > 0$, for every $A \in \Gamma$ and $\mathbb{I}(B) = 0$ for every $B \in \Delta$.

Identity may now be defined in a standard way:

Definition 5 (*Identity*) $(a =_{LP} b) \quad =_{df} \quad \forall P(Pa \equiv Pb)$

1.3 Assessment

Priest's characterization of identity in second-order LP has the effect of "relaxing" some of the properties of classical identity. Consider the following toy model, where for all functions $f \in D_2$, $f(a) = f(b) = f(c)$ except for a function f^* that $f^*(a) = 1$, $f^*(b) = \frac{1}{2}$ and $f^*(c) = 0$. In Priest's dialetheist reading of the semantics, this corresponds to a situation where all the properties are shared similarly by a, b

Table 1 Dialetheist interpretation of $\frac{1}{2}$

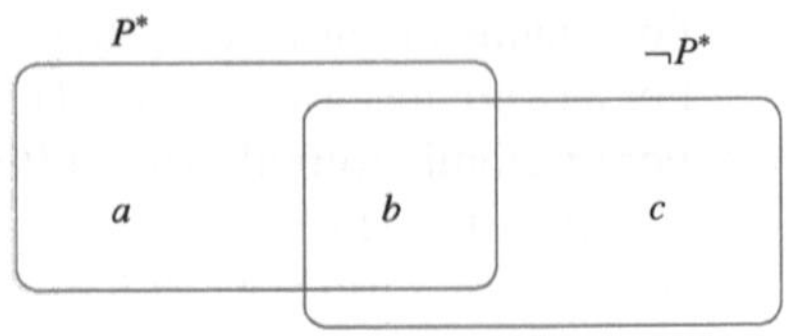

and c except for a property P^* that a has (but not its complement), b has (just as its complement) and c lacks (but does have its complement), see Table 1.

N

Identity is both reflexive and symmetric (as it should be). The non-transitivity of identity is inherited from the non-transitivity of LP's material conditional. In the case at hand, for all substitution of P by a predicate interpreted by a function in D2: $Pa \equiv Pb$ and $Pb \equiv Pc$ although it is not the case that for all substitutions of P $(Pa \equiv Pc)$. A second feature is inherited from LP's material conditional. LP's material conditional is not detachable, in the sense that *modus ponens* can fail. This leads, in the case of identity, to a failure of *substitutivity*. The toy model above is a countermodel showing that $b =_{LP} c,\ P^*b \nvDash_{LP} P^*c$.

Although we find the general approach reasonable, we think the last feature of Priest's proposal is not particularly pleasing. Think of the definition of identity: that is based on the Leibnizian idea according to which identity is a matter of sharing all properties. But the failure of substitutivity clashes with the spirit of the Leibnizian idea. It might be objected that the failure of transitivity is a particular case of failure of substitutivity. That's true, but identity has been defined as sharing all "relevant" properties (note that D_2 need not equal $\{1, \frac{1}{2}, 0\}^{D_1}$). Substitutivity should work at least for "relevant" properties.

In the next section we develop two notions of identity built on ideas close to Priest's. Our first notion of identity is non-transitive but substitutivity works. That's already, we think, an improvement over Priest's notion. Second, we develop a notion of identity that is fully transitive. Despite its classicality, this second notion of identity is sensitive to expressions of (in)definiteness; we want to argue, against the widespread opinion, that a transitive notion of identity is compatible with a metaphysical reading of indefiniteness.

2 Two Notions of Tolerant Identity

2.1 *Second-Order ST*

In Ripley [9] and Cobreros et al. [2, 3] we investigate a logic that retains some affinities with LP while remaining faithful to classical logic in many respects. The semantics for our logic ST (as we shall call it) is exactly that of LP above. The difference concerns the definition of logical consequence:

Definition 6 We say that $\Gamma \models^{ST} \Delta$ iff there is no MV-model M such that $\mathbb{I}(A) = 1$, for every $A \in \Gamma$ and $\mathbb{I}(B) = 0$ for every $B \in \Delta$.

The logic ST sets different standards for satisfaction in premises and in conclusions. A "good" premise (a premise good enough to produce a sound argument) is one that takes value 1. A "good" conclusion, on the other hand (a conclusion that is not false enough to produce a counterexample) is one that takes value greater than 0. This definition might be viewed as setting a *permissive* relation of logical consequence (see [3], Sect. 2.2). For the classical vocabulary (no expressions for indefiniteness or the like) the logic LP coincides with classical logic in its theorems: A is classically valid just in case it is LP-valid. A striking feature of ST is that, for the classical vocabulary, the logic is *fully classical*: Δ is a classical consequence of Γ just in case Δ is an ST-consequence of Γ. However, the logic is sensitive to expressions that do not belong to a purely classical first-order vocabulary (in Cobreros et al. [1] we investigate this logic in connection to the sorites paradox where similarity relations are around; in Cobreros et al. [2] we investigate ST-logic in combination with a transparent truth predicate and self-reference). When non-classical expressions are around, the logic ST might lead to failures of transitivity, thereby blocking inferences that would otherwise trivialize the theory.

The definition of ST carries over from MV to MV_2-models to provide a second order version of this logic:

Definition 7 We say that $\Gamma \models_2^{ST} \Delta$ iff there is no MV_2-model M such that $\mathbb{I}(A) = 1$, for every $A \in \Gamma$ and $\mathbb{I}(B) = 0$ for every $B \in \Delta$.

For all that was pointed out above it can be seen that second-order ST is equivalent to (a version of) second-order classical logic.

2.2 *Tolerant Identity, First Try*

Our first notion of tolerant identity is defined making use of the machinery of MV_2-models above:

Definition 8 ($Tol\,id.\ 1^{st}$). $\mathbb{I}(a \approx b) = 1$ just in case for every $f \in D_2$, $|f(a) - f(b)| < 1$

The expression $|f(a) - f(b)| < 1$ states that a and b are similar with respect to property f. Thus, this definition states that similarity in all properties is sufficient in order to have a corresponding statement of identity good enough to produce a sound argument.

It's easy to see that '$\approx$' is both reflexive and symmetric; and a toy model as the one employed above suffices to show that the relation is not transitive. Recall, that is a model where for all functions $f \in D_2$, $f(a) = f(b) = f(c)$ except for a function f^* that $f^*(a) = 1$, $f^*(b) = \frac{1}{2}$ and $f^*(c) = 0$. The ST-reading of the semantics is different from the LP reading, though. Now values are relative to the position of

Table 2 *ST* interpretation of $\frac{1}{2}$

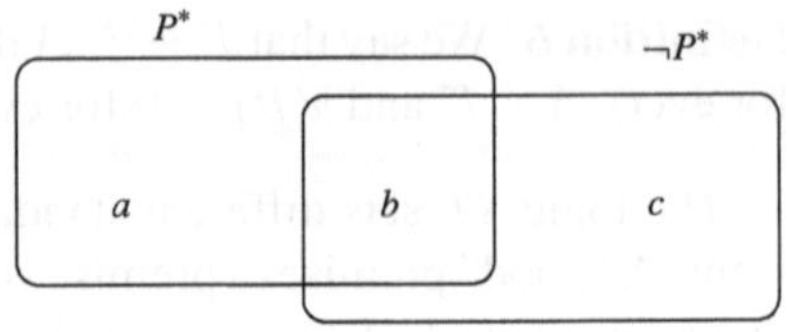

corresponding sentences in premises or conclusions of an argument. In Table 2, the upper oval indicates a "good conclusion" (cannot produce a counterexample) and lower oval a "bad premise" (cannot produce a sound argument).

This notion of identity does retain substitutivity in the sense that the following properties hold:

(Subst1) $\models_2^{ST} \forall x \forall y \forall P((Px \wedge x \approx y) \supset Py)$
(Subst2) $Px, x \approx y \models_2^{ST} Py$

Note that in order for (Subst1) to fail, the conditional should take value 0. This occurs just in case the antecedent is 1 and the consequent is 0. But if '$Px \wedge x \approx y$' takes value 1, then Py should take at least value $\frac{1}{2}$. Similarly for (Subst2): for that inference to fail there must be a model where premises are 1 and conclusion is 0. But the value 1 of '$Px, x \approx y$' guarantees a value greater than 0 for Py.

2.3 Tolerant Identity, Second Try

Our second definition of identity directly mirrors Priest's strategy but within the (second-order) ST-logic.

Definition 9 ($Tol\,id.2^{nd}$). $(a = b) \quad =_{df} \quad \forall P(Pa \equiv Pb)$

Despite the affinities in the semantics, the classicality of ST's material conditional makes this notion of identity fully transitive. This notion of identity is, however, sensitive to non-classical expressions. Consider Priest's motorbike once again. At each stage, the resulting motorbike is similar in all its properties to the previous one. That is, for each of the stages a_n of the motorbike we have $a_n \approx a_{n+1}$ ('$\approx$' understood as defined in the previous section). Although the notion of identity introduced in Definition 9 is classical, it is tolerant in connection to the similarity relation introduced in Definition 8. That is, the following *tolerance principles for identity* hold in ST:

TPI1 $\models_2^{ST} \forall x \forall y((PM = x \wedge x \approx y) \supset PM = y)$
TPI2 $a \approx b,\ a = PM \models_2^{ST} b = PM$

Table 3 Evans' argument

1	$\nabla(b = a)$	(Assumption)
2	$\lambda x[\nabla(x = a)]b$	(From 1, by abstraction)
3	$\neg\nabla(a = a)$	(Assumption, since $a = a$ is a logical truth!)
4	$\neg\lambda x[\nabla(x = a)]a$	(From 3, by abstraction)
5	$\neg(a = b)$	(From 2, 4, by LL and Contrap)
6	$\neg\nabla(a = b)$	(Assuming premises are definite)

In short, although identity is fully classical, it is a tolerant relation (unlike standard classical identity). This fact can be used to explain our intuitions about non-sharp transitions in cases like that of Priest's motorbike. At the same time, the non-transitivity of the ST-logic is what prevents the unwelcome conclusion of the sorites paradox.

3 Transitivity and Metaphysical Indeterminacy

In the previous section we argued that Priest's approach to identity can be refined. First, the failure of substitutivity deprives the Leibnizian definition of identity of its intended force. Within the ST-logic, we can define a non-transitive notion of identity for which substitutivity works. Second, within the ST-logic, we can define a notion of identity that is fully transitive (and, naturally, for which substitutivity works) but that is tolerant, and so it makes still room for indeterminacy. In this section we want to argue that the indeterminacy associated to this notion of identity need not be understood in a purely semantic way. Thus, against a widespread opinion, we argue that transitivity of identity and metaphysical indeterminacy are compatible.

In order to show this, we consider Evans' famous argument (in [4]), under Lewis' interpretation (in [7]). Evans' argument is a *reductio* from the assumption of a true statement of indefiniteness of identity ('∇A' means 'it is indefinite whether A'). See Evans' arguement in Table 3.

Naturally, this argument must be fallacious, since it is perfectly agreed that there might be indefinite identity statements. Lewis' interpretation of Evans' argument is that while the defender of indeterminacy as semantic can easily point out where the fallacy lies, the same is not the case for the defender of indeterminacy as metaphysical. In particular, one can say that the steps from 1 to 2 and 3 to 4 are not valid, in much the same way as the inference from the true statement,

'It is contingent whether the number of planets is eight'

does not entail the false statement,

'the number of planets is such that it is contingent whether it is eight'.

Now this analogy makes perfect sense if indeterminacy is understood in terms of a variation of the denotation of a term across precisifications (that is the supervaluationist reading). In that case, 'contingency' and 'indeterminacy' are formally identical and the mentioned inference is not valid. But for the defender of indeterminacy as metaphysical, indeterminacy cannot be explained in terms of variation of the denotation across anything. The terms a and b in the argument "rigidly denote" (to follow the modal analogy) an object that is intrinsically vague.

We take Evans' argument (under Lewis' interpretation) as a criterion for the availability of a metaphysical reading of indeterminacy associated to identity. Given notions of identity and of indeterminacy, if the only way to block the argument is the invalidity of abstraction into the scope of the ∇ operator, then that notion of indeterminacy has just a semantic reading.

Let now definiteness ('it is definite *that*') be defined as follows:

Definition 10 (*Definiteness*)

$$\mathbb{I}(\mathsf{D}(a = b)) = \begin{cases} 1 \text{ if for all } f \in D_2 \quad |f(a) - f(b)| = 0 \\ 0 \text{ otherwise} \end{cases}$$

Indefiniteness ('it is indefinite *whether*') as expressed by '∇' can be defined thus:

$$\nabla(a = b) = \neg\mathsf{D}(a = b) \wedge \neg\mathsf{D}\neg(a = b),$$

and consider again Evans' argument. Each step in the argument is ST-valid. However, (5) is only tolerantly true and (6) is not even tolerantly true. Thus, though each step is valid we cannot validly chain premises in this case.

4 Conclusion and Outlook

Priest argued that to account for substantial change, one must admit that identity is non-transitive. Making use of the logic LP, he defines such a notion of identity, which also fails to satisfy substitutivity. Making use of our alternative logic ST, we have shown that we can not only define a non-transitive notion of identity that preserves substitutivity, but also a notion of identity that is transitive. The latter is particularly interesting, because even though transitive, it is still a tolerant relation which allows for substantial change. Addressing Evans' argument, we have also shown that the transitivity of identity is compatible with ontological vagueness.

In this chapter we have focussed on logical issues. However, we believe that our proposed analyses of identity have interesting ontological implications. We mentioned already the issues of substantial change and ontological vagueness. But both deserve more extensive discussion: how is substantial change compatible with the transitivity of identity from a conceptual point of view, and what does it mean to be a vague object? Identity is crucial for counting, but what is the consequence for counting when our notions of identity are used? Last but not least, there is Geach's

[5] notion of 'relative identity' and Unger's [10] problem of the Many. We feel that for both a fresh perspective becomes available when use is made of the notions introduced in this chapter. We hope to address these issues in a subsequent chapter.

References

1. Cobreros, P., Egre, P., Ripley, D., & van Rooij, R. (2012). Tolerant, classical, strict. *Journal of Philosophical Logic*, *41*(2), 347–85.
2. Cobreros, P., Egré, P., Ripley, D., & van Rooij, R. (2014). Reaching transparent truth. *Mind* (forthcoming).
3. Cobreros, P., Egré, P., Ripley, D., & van Rooij, R. (2014). Vagueness, truth and permissive consequence. In T. Achourioti, H. Galinon, K. Fujimoto & J. Martínez-Fernández (Eds.), *Unifying the Philosophy of Truth*. Springer (forthcoming).
4. Evans, G. (1978). Can there be vague objects? *Analysis*, *38*(4), 208.
5. Geach, P. T. (1980). *Reference and generality* (3rd ed.). Ithaca: Cornell University Press.
6. Lewis, D. (1986). *On the plurality of worlds*. Oxford: Blackwell.
7. Lewis, D. (1988). Vague identity: Evans misunderstood. *Analysis*, *48*(3), 128–30.
8. Priest, G. (2010). Non-transitive identity. In R. Dietz & S. Moruzzi (Eds.), *Cuts and Clouds*. Oxford: Oxford University Press.
9. Ripley, D. (2012). Conservatively extending classical logic with transparent truth. *The Review of Symbolic Logic*, *5*(2), 354–78.
10. Unger, P. (1980). The problem of the many. *Midwest Studies in Philosophy*, *5*(1), 411–67.

How to Unify Russellian and Strawsonian Definite Descriptions

Marie Duží

Abstract In this paper I will deal with ambiguities in natural language exemplifying the difference between *topic* and *focus articulation* within a sentence. I will show that whereas articulating the topic of a sentence activates a presupposition, articulating the focus frequently yields merely an entailment. Based on analysis of topic-focus articulation, I propose a solution to the almost hundred-year old dispute over Strawsonian versus Russellian definite descriptions. The point of departure is that sentences of the form 'The *F* is a *G*' are ambiguous. Their ambiguity stems from different topic-focus articulations of such sentences. Russell and Strawson took themselves to be at loggerheads, whereas, in fact, they spoke at cross purposes. My novel contribution advances the research into definite descriptions by pointing out how progress has been hampered by a false dilemma and *how to move beyond that dilemma*. The point is this. If 'the *F*' is the topic phrase then this description occurs with *de re* supposition and Strawson's analysis appears to be what is wanted. On this reading the sentence *presupposes* the existence of the descriptum of 'the *F*'. The other option is '*G*' occurring as topic and 'the *F*' as focus. This reading corresponds to Donnellan's attributive use of 'the *F*' and the description occurs with *de dicto* supposition. On this reading the Russellian analysis gets the truth-conditions of the sentence right. The existence of a unique *F* is merely entailed. This paper demonstrates how to unify these disparate insights into one coherent theory of definite descriptions.

Keywords Definite descriptions · Presupposition versus entailment · Topic-focus · Transparent intensional logic

---

M. Duží (✉)
Department of Computer Science, VSB-Technical University of Ostrava, 17. listopadu 15, 70833 Ostrava, Czech Republic
e-mail: marie.duzi@vsb.cz

R. Ciuni et al. (eds.), *Recent Trends in Philosophical Logic*,
Trends in Logic 41, DOI: 10.1007/978-3-319-06080-4_7,

1 Introduction

Natural language has features not found in logically perfect artificial languages. One such feature is *redundancy*; another feature is its converse, namely *ambiguity*. In this paper I will deal with the sort of ambiguity that is pivoted on whether the *topic* or the *focus* of a sentence is highlighted. For instance, 'John only introduced Bill to Sue' lends itself to two different kinds of construal: 'John did not introduce other people to Sue except for Bill' and 'The only person Bill was introduced to by John was Sue'.[1] There are two sentences whose semantics, logical properties and logical consequences only partially overlap.

Based on analysis of sentences that differ as for their topic-focus articulation I propose a solution to the almost hundred-year old dispute over Strawsonian versus Russellian definite descriptions.[2] The point of departure is that sentences of the form 'The *F* is a *G*' are systematically ambiguous.[3] Their ambiguity is, in my view, not rooted in a shift of meaning of the definite description 'the *F*'. Rather the ambiguity stems from different *topic-focus articulations* of such sentences. My analysis assumes that whereas articulating the topic of a sentence activates a pre- supposition, articulating the focus frequently yields merely an entailment.[4] The point is this. If 'the *F*' is the topic phrase then this description occurs with *de re* supposition and Strawson's analysis appears to be what is wanted. On this reading that corresponds to Donnellan's *referential use* of 'the *F*' the sentence *presupposes* the existence of the descriptum of 'the *F*'. The other option is '*G*' occurring as topic and 'the *F*' as focus. This reading corresponds to Donnellan's *attributive* use of 'the *F*' and the description occurs with *de dicto* supposition. On this reading the Russellian analysis gets the truth-conditions of the sentence right. The existence of a unique *F* is merely entailed.

The received view still tends to be that there is room for at most one of the two positions, since they are deemed incompatible. But there is no incompatibility between Strawson's and Russell's positions, because they simply do not talk about one and the same meaning of the sentence 'The King of France is bald'. My novel *contribution* is to point out this *ambiguity* which yielded the false dilemma. Russell argued for attributive use of 'the King of France' whereas Strawson for its referential use. In this paper I will propose a logical analysis of both Russellian and Strawsonian reading of sentences of the form 'The *F* is a *G*'.

Tichý's Transparent Intensional Logic (TIL) will serve as background theory throughout my exposition.[5] Tichý's TIL was developed simultaneously with Montague's IL (Intensional Logic). The technical tools of disambiguation will be familiar

[1] See Hajičová [7].

[2] See for instance Refs. [2, 13–16, 18, 21].

[3] The sentence that triggered the dispute was 'The King of France is bald'.

[4] This assumption is based on [7], and supported by other linguists as well. See, for instance [6], Gundel and Fretheim, in press, http://www.sfu.ca/~hedberg/gundel-fretheim.pdf, and [17, esp. p. 173ff].

[5] For details on TIL, see, in particular [5, 19, 20].

from IL, with two exceptions. One is that we λ-bind separate variables $w, w_1, \ldots, w_n$ ranging over possible worlds and $t, t_1, \ldots, t_n$ ranging over times. This dual binding is tantamount to *explicit intensionalization* and *temporalization*. The other exception is that *functional application* is the logic both of extensionalization of intensions (functions from possible worlds) and of predication.[6] Application is symbolized by square brackets, '[...]'. Intensions are extensionalized by applying them to worlds and times, as in $[[Intension\ w]\ t]$, abbreviated by subscripted terms for world and time variables: $Intension_{wt}$ is the extension of the generic intension *Intension* at $\langle w, t\rangle$. Thus, for instance, the extensionalization of a property yields a set (possibly an empty one), and the extensionalization of a proposition yields a truth-value (or no value at all). A general objection to IL is that it fails to accommodate *hyperintensionality*, as indeed any formal logic interpreted set-theoretically is bound to unless a domain of primitive hyperintensions is added to the frame. Any theory of natural-language analysis needs a hyperintensional semantics in order to crack the hard nuts of natural language semantics. In global terms, divested of its hyperintensional procedural semantics TIL is an anticontextualist (i.e., transparent), explicitly intensional modification of IL. With its hyperintensional procedural semantics added back on, TIL rises above the model-theoretic paradigm and joins instead the paradigm of hyperintensional logic and structured meanings.[7]

The rest of the paper is organized as follows. Section 2 is a brief summary of the bones of contention between Russsellian and Strawsonian conceptions of definite descriptions. The relevant foundations of TIL are introduced in Sect. 3. Finally, in Sect. 4 I propose my unification of elements drawn from Strawsonian and Russellian theories of definite descriptions.

2 Russell Versus Strawson on Definite Descriptions

There is a substantial difference between proper names and definite descriptions. This distinction is of crucial importance due to their vastly different logical behaviour. Independently of any particular theory of proper names, it should be granted that a *proper* proper name (as opposed to a definite description grammatically masquerading as a proper name) is a rigid designator of a numerically particular individual. On the other hand, a definite *description* like, for instance, 'the King of France', 'the highest mountain on earth', 'the first man to run 100 m in under 9 seconds', etc., offers an *empirical criterion* that enables us to establish which individual, if any, satisfies the criterion in a particular state of affairs.

The contemporary discussion of the distinction between names and descriptions was triggered by [14]. Russell's key idea is the proposal that a sentence like

(1) 'The *F* is a *G*.'

containing a definite description 'the *F*' is understood to have, in the final analysis, the logical form

---

[6] For details, see Jespersen [8].

[7] For a detailed critical comparison of TIL and IL, see [5, §2.4.5].

(1′) $\exists x(Fx \wedge \forall y(Fy \supset x = y) \wedge Gy)$

rather than the logical form $G(\iota xFx)$.

Though Russell's quantificational theory remains to this day a strong rival of referential theories, it has received its fair share of criticism. Russell's opponents claim that he simply gets the truth-conditions wrong in important cases of using descriptions when there is no such thing as the unique *F*.[8]

This criticism was launched by Strawson who in 1950 objected that Russell's theory predicts the wrong truth-conditions for sentences like 'The present King of France is bald'. According to Russell's analysis, this sentence is false. In Strawson's view, the sentence can be neither true nor false whenever there is no unique King of France. Obviously, in such a state of affairs the sentence is not true. However, if the sentence were false then its negation, 'The King of France is *not* bald', would be true, which entails that there *is* a unique King of France, contrary to the assumption that there is none. Strawson holds that sentences like these *not only entail*, but also *presuppose*, the existence of a unique King of France. If 'the present King of France' fails to refer, then the presupposition is not satisfied and the sentence fails to have a truth value.[9]

Russell, in response to Strawson's criticism, argues that, despite Strawson's protests, the sentence is in fact false:

> Suppose, for example, that in some country there was a law that no person could hold public office if he considered it false that the Ruler of the Universe is wise. I think an avowed atheist who took advantage of Mr. Strawson's doctrine to say that he did not hold this proposition false would be regarded as a somewhat shifty character [15].

Donnellan [2] observes that there is a sense in which Strawson and Russell are both right, and both wrong, about the proper analysis of definite descriptions, because definite descriptions can be used in two different ways. On a so-called *attributive use*, a sentence of the form 'The *F* is a *G*' is used to express a proposition equivalent to 'Whatever is uniquely *F* is a *G*'. Alternatively, on a *referential use*, a sentence of the form 'The *F* is a *G*' is used to pick out a specific individual, *a*, and to say of *a* that *a* is a *G*. Donnellan suggests that Russell's quantificational account of definite descriptions might capture attributive uses, but that it does not work for referential uses. Ludlow in 2007 interprets Donnellan as arguing that in some cases descriptions are Russellian and in other cases Strawsonian.

Kripke [11] responds to Donnellan by arguing that the Russellian account of definite descriptions can, by itself, account for both referential and attributive uses, and that the difference between the two cases is entirely a matter of pragmatics. Neale [13] supports Russell's view by collecting a number of cases in which intuitions about truth conditions clearly do not support Strawson's view. On the other hand, a number

---

[8] Besides, many hold against Russell's translation of atomic sentences like 'The *F* is a *G*' into the molecular form 'There is at least one *F* and at most one thing is an *F* and that thing is a *G*', because Russell disregards the standard constraint that there must be a fair amount of structural similarity between analysandum and analysans.

[9] Nevertheless, for Strawson, *sentences* are meaningful in and of themselves, independently of empirical facts like the contingent non-existence of the King of France.

of linguists have recently come to Strawson's defence on this matter. See Ludlow [12] for a detailed survey of the arguments supporting Strawson's view and arguments supporting Russell's. Here it might suffice to point out that Strawson's concerns have not delivered a knock-out blow to Russell's theory of descriptions, and so this topic remains very much alive. von Fintel [21], for instance, argues that every sentence containing a definite description 'the F' comes with the existential presupposition that there be a unique F.

In this paper I am not going to take into account Kripke's pragmatic factors like the intentions of a speaker, for they are irrelevant to a *logical* semantic theory. So I am disregarding Donnellan's troublesome notion of having somebody in mind. Instead, I will propose a *logical analysis* of sentences of the form 'The F is a G'. What I want to show is this. First, definite descriptions are not deprived of a self-contained meaning and they denote one and the same entity in any context. Thus they are never Russellian. Second, Russells insight that a definite description 'the F' does not denote a definite individual is spot-on. According to TIL, 'the F' denotes a *condition* to be contingently satisfied by the individual (if any) that happens to be the F. I will explicate such conditions in terms of possible-world intensions, *viz.* as individual roles or offices to be occupied by at most one individual per world/time pair. Third, I am going to show that Donnellan is right in holding that sentences of the form 'The F is a G' are systematically ambiguous. However, their ambiguity does not concern a shift of meaning of the definite description 'the F', as Fregean or other theories maintain. Instead the ambiguity concerns different *topic-focus* articulations of these sentences.

There are two options. The description 'the F' may occur as the topic of a sentence and property G (the focus) is predicated of the topic. This case corresponds to Donnellan's *referential use*. Using medieval terminology I will say that 'the F' occurs with *de re supposition.* The other option is 'G' occurring as topic and 'the F' as focus. This reading corresponds to Donnellan's *attributive* use of 'the F' and the description occurs with *de dicto* supposition. Consequently, and crucially, such sentences are ambiguous between a *de dicto* and a *de re* reading. On their *de re* reading they *presuppose* the existence of a unique F. Thus Strawson's analysis appears to be adequate for *de re* cases. On their *de dicto* reading they have the truth-conditions as specified by the Russellian analysis. They do not presuppose, but only entail, the existence of a unique F. However, the Russellian analysis, though being equivalent to the one I am going to propose, is not an adequate *literal* analysis of *de dicto* readings.

I am going to bring out the *semantic* nature of the topic-focus difference by means of a literal logical analysis. As a result, I will be furnishing sentences differing only as for their topic-focus articulation with different structured meanings producing different possible-world propositions.[10] Since our logic is a hyperintensional logic of *partial functions*, I am able to analyse sentences with presuppositions in a both natural and principled manner. It means that I associate them with hyperpropositions,

---

[10] For details on structured meanings, see [4, 10] for a survey.

which in TIL are abstract logical procedures that produce partial possible-world propositions, which occasionally yield truth-value gaps.[11]

We need to work with properly partial functions and propositions with truth-value gaps. On Strawsonian reading the sentence 'The King of France is bald' talks about the office of the King of France (topic) ascribing to the individual (if any) that occupies this office the property of being bald (focus). Thus it is presupposed that the King of France exist, i.e., that the office be occupied. If the office is vacant the proposition denoted by the sentence lacks a truth-value. On our approach this does not mean that the sentence is meaningless. The sentence has a sense, namely an instruction how in any possible world w at any time t to execute the procedure of evaluating its truth-conditions. Only if we evaluate these conditions in such a state-of-affairs where there is no King of France does the process of evaluation yield a truth-value gap.

3 Foundations of TIL

Formally, TIL is an extensional logic of hyperintensions based on the partial, typed λ-calculus enriched with a ramified type structure to accommodate hyperintensions. The syntax of TIL is the familiar one of the λ-calculus, with the addition of a hyperintension called Trivialization (symbolized by a superscripted nought). The semantics is a *procedural* (as opposed to denotational) one. Thus, functional application, in TIL, is not the result of applying a function to an argument, but instead the very *procedure* of applying function to argument; and functional abstraction, in TIL, is not the result of forming a function, but instead the very *procedure* of sorting two domains of entities into functional arguments and values, respectively. The TIL concept of procedurally construed hyperintensions is *construction*. The three definitions below constitute the logical heart of TIL.

Definition 1 *(Types of order 1.)* Let B be a *base*, where a base is a collection of pair-wise disjoint, non-empty sets. Then:

(i) Every member of B is an elementary *type of order 1 over B*.
(ii) Let $\alpha, \beta_1, \ldots, \beta_m$ $(m > 0)$ be types of order 1 over B. Then the collection $(\alpha\ \beta_1 \ldots \beta_m)$ of all m-ary partial mappings from $\beta_1 \times \cdots \times \beta_m$ into α is a functional type of order 1 over B.
(iii) Nothing is a *type of order 1 over B* unless it so follows from (i) and (ii). □

Remark For the purposes of natural-language analysis, we are currently assuming the following base of ground types, each of which is part of the ontological commitments of TIL:

o: the set of truth-values $\{\mathbf{T}, \mathbf{F}\}$;
ι: the set of individuals (a constant universe of discourse);
τ: the set of real numbers (doubling as temporal continuum);

[11] For an introduction to the notion of hyperproposition, see [9].

ω: the set of logically possible worlds (the logical space).

Constructions construct objects of appropriate types dependently on *valuation* of variables; they *v-construct*, where v is the parameter of valuation. With the difference that we construe variables as extra-linguistic objects and not as expressions, our theory of variables is otherwise identical to Tarski's. Thus, in TIL variables construct objects of the respective types dependently on valuation in the following way. For each type α there are countably infinitely many variables $x_1, x_2, \ldots$. The members of α (unless α is a singleton) can be organised in infinitely many infinite sequences. Let the sequences be given (as one is allowed to assume in a realist semantics). The valuation v takes a sequence $\langle s_1, s_2, \ldots \rangle$ and assigns s_1 to the variable x_1, s_2 to the variable x_2; and so on.

When X is an object of any type (including a construction), the Trivialization of X, denoted '${}^0X$', constructs X without the mediation of any other constructions. ${}^0X$ is the unique atomic construction of X that does not depend on valuation: it is a primitive, non-perspectival mode of presentation of X. The other constructions are *compound*, as they consist of other constituents apart from themselves. These are *Composition* and *Closure*. Composition is the procedure of applying a function f to an argument a to obtain the value (if any) of f at a. Closure is the procedure of constructing a function by abstracting over variables; i.e., the procedure of abstracting, or extracting, a function from a context, as when abstracting $\lambda x(\phi x)$ from $\phi(a)$.[12]

Definition 2 *(construction)*

(i) The *variable* x is a *construction* that constructs an object O of the respective type dependently on a valuation v: x *v-constructs* O.
(ii) *Trivialization:* Where X is an object whatsoever (an extension, an intension or a *construction*), ${}^0X$ is the *construction Trivialization*. It constructs X without any change in X.
(iii) The *Composition* $[X\ Y_1 \ldots Y_m]$ is the following construction. If X *v-constructs* a function f of type $(\alpha\ \beta_1 \ldots \beta_m)$, and $Y_1, \ldots, Y_m$ *v-construct* entities $B_1, \ldots, B_m$ of types $\beta_1, \ldots, \beta_m$, respectively, then the *Composition* $[X\ Y_1 \ldots Y_m]$ *v-constructs* the value (an entity, if any, of type α) of f on the tuple argument $\langle B_1, \ldots, B_m \rangle$. Otherwise the *Composition* $[X\ Y_1 \ldots Y_m]$ does not *v-construct* anything and so is *v-improper*.
(iv) The *Closure* $[\lambda x_1 \ldots x_m\ Y]$ is the following construction. Let $x_1, x_2, \ldots, x_m$ be pair-wise distinct variables v-constructing entities of types $\beta_1, \ldots, \beta_m$ and Y a construction v-constructing an α-entity. Then $[\lambda x_1 \ldots x_m\ Y]$ is the *construction λ-Closure* (or *Closure*). It *v-constructs* the following function f of the type $(\alpha\beta_1 \ldots \beta_m)$. Let $v(B_1/x_1, \ldots, B_m/x_m)$ be a valuation identical with v at least up to assigning objects $B_1/\beta_1, \ldots, B_m/\beta_m$ to variables $x_1, \ldots, x_m$. If Y is $v(B_1/x_1, \ldots, B_m/x_m)$-improper (see iii), then f is undefined on $\langle B_1, \ldots, B_m \rangle$.

[12] There are two other compound construction; Execution and Double Execution. Since I do not need them in this paper, they are not incorporated in Definition 2.

Otherwise the value of f on $\langle B_1, \ldots, B_m\rangle$ is the α-entity $v(B_1/x_1, \ldots, B_m/x_m)$-constructed by Y.

(v) Nothing is a *Construction*, unless it follows from (i) through (iv). □

The definition of the ramified hierarchy of types decomposes into three parts. Firstly, simple types of order 1, which were already defined by definition 1. Secondly, constructions of order n, and thirdly, types of order $n + 1$.

Definition 3 *(Ramified Hierarchy of Types)*

T1 *(types of order 1)*. See Definition 1.

Cn *(constructions of order n)*

(i) Let x be a variable ranging over a type of order n. Then x *is a construction of order n over B*.

(ii) Let X be a member of a type of order n. Then ${}^0X$, ${}^1X$, ${}^2X$ are *constructions of order n over B*.

(iii) Let $X, X_1, \ldots, X_m$ $(m > 0)$ be constructions of order n over B. Then $[X\ X_1 \ldots X_m]$ is a *construction of order n over B*.

(iv) Let $x_1, \ldots, x_m, X$ $(m > 0)$ be constructions of order n over B. Then $[\lambda x_1 \ldots x_m\ X]$ is a *construction of order n over B*.

(v) Nothing is a *construction of order n over B* unless it so follows from **Cn** (i)–(iv).

Tn+1 (types of order $n + 1$). Let $*_n$ be the collection of all constructions of order n over B. Then

(i) $*_n$ and every type of order n are *types of order* $n + 1$.

(ii) If $m > 0$ and $\alpha, \beta_1, \ldots, \beta_m$ are types of order $n + 1$ over B,then $(\alpha\ \beta_1 \ldots \beta_m)$ (see T_1(ii)) is a *type of order* $n + 1$ *over B*.

(iii) Nothing is a *type of order* $n + 1$ *over B* unless it so follows from (i) and (ii). □

Empirical languages incorporate an element of *contingency* that non-empirical ones lack. Empirical expressions denote *empirical conditions* that may or may not be satisfied at some empirical index of evaluation. We model these empirical conditions as *possible-world intensions*. Intensions are entities of type $(\beta\omega)$: mappings from possible worlds to an arbitrary type β. The type β is frequently the type of the *chronology* of α-objects, i.e. a mapping of type $(\alpha\tau)$. Thus α-intensions are frequently functions of type $((\alpha\tau)\omega)$, abbreviated as '$\alpha_{\tau\omega}$'. I shall typically say that an index of evaluation is a world/time pair $\langle w, t\rangle$. *Extensional entities* are entities of some type α where $\alpha \neq (\beta\omega)$ for any type β.

Examples of frequently used intensions are: *propositions* of type $o_{\tau\omega}$, *properties of individuals* of type $(o\iota)_{\tau\omega}$, binary *relations-in-intension between individuals* of type $(o\iota\iota)_{\tau\omega}$, *individual offices* of type $\iota_{\tau\omega}$. Thus individual offices are simply partial functions which, relative to a world/time pair $\langle w, t\rangle$, return at most one individual as value.

Our *explicit intensionalization and temporalization* enables us to encode constructions of possible-world intensions, by means of terms for possible-world variables and times, directly in the logical syntax. Where w ranges over ω and t over τ, the following general logical form characterizes the logical syntax of constructions of intensions: $\lambda w \lambda t[\ldots w \ldots t \ldots]$. For instance, if *King_of* is a function of type $(\iota\iota)_{\tau\omega}$ and *France* an individual of type ι, the office of the King of France is constructed like this: $\lambda w \lambda t[{}^0King\_of_{wt}\ {}^0France]$.

Logical objects like *truth-functions* and *quantifiers* are extensional: $\wedge$ (conjunction), $\vee$ (disjunction) and $\supset$ (implication) are of type (οοο), and $\neg$ (negation) of type (οο). *Quantifiers* $\forall^\alpha$, $\exists^\alpha$ are type-theoretically polymorphous, total functions of type $(o(o\alpha))$, for an arbitrary type α, defined as follows. The *universal quantifier* $\forall^\alpha$ is a function that associates a class A of α-elements with **T** if A contains all elements of the type α, otherwise with **F**. The existential quantifier $\exists^\alpha$ is a function that associates a class A of α-elements with **T** if A is a non-empty class, otherwise with **F**.

Below all type indications will be provided outside the formulae in order not to clutter the notation. Furthermore, 'X/α' means that an object X is (a member) of type α. '$X \rightarrow_v \alpha$' means that the type of the object *valuation*-constructed by X is α. Throughout, it holds that the variables $w \rightarrow_v \omega$ and $t \rightarrow_v \tau$. If $C \rightarrow_v \alpha_{\tau\omega}$, then the frequently used Composition $[[C\,w]\,t]$, which is the intensional descent (a.k.a. extensionalization) of the α-intension v-constructed by C, will be encoded as 'C_{wt}'. When using constructions of truth-functions, we often omit Trivialization and use infix notation to conform to standard notation in the interest of better readability. Also when using constructions of identities of α-entities, $=_\alpha /(o\alpha\alpha)$, we omit Trivialization, the type subscript, and use infix notion when no confusion can arise.

We invariably furnish expressions with procedural structured meanings, which are explicated as TIL constructions. The analysis of an unambiguous empirical sentence thus consists in discovering the logical construction encoded by a given sentence. The *TIL method of analysis* consists of three steps:

(1) *Type-theoretical analysis*, i.e., assigning types to the objects that receive mention in the analysed sentence.
(2) *Type-theoretical synthesis*, i.e., combining the constructions of the objects ad (1) in order to construct the proposition of type $o_{\tau\omega}$ denoted by the whole sentence.
(3) *Type-theoretical checking*, i.e. checking whether the proposed analysans is type-theoretically coherent.

To illustrate the method, we analyse the stock example 'The King of France is bald' *à la* Strawson.

First, type-theoretical analysis. The sentence mentions these objects. *King_of*/$(\iota\iota)_{\tau\omega}$ is an empirical function that dependently on $\langle w, t\rangle$-pairs assigns to one individual (a country) another individual (its king); *France*/ι; *King_of_France*/$\iota_{\tau\omega}$; *Bald*/$(o\iota)_{\tau\omega}$.

For the sake of simplicity, I will demonstrate the steps (2) and (3) simultaneously. In the second step we combine the *constructions* of the objects obtained in the first step in order to construct the proposition (of type $o_{\tau\omega}$) denoted by the whole sentence. Since we intend to arrive at the *literal* analysis of the sentence, the objects denoted by the semantically simple expressions are constructed by their Trivializations: ${}^0King\_of$, ${}^0France$, ${}^0Bald$. In order to construct the office *King_of_France*, we have to combine $^0King\_of$ and ^0France. The function $King\_of$ must be extensionalized first *via* the Composition $^0King\_of_{wt} \rightarrow_v (\iota\iota)$, and the result is then applied to *France*; we get $[^0King\_of_{wt}\ ^0France] \rightarrow_v \iota$. Abstracting over the values of w and t we obtain the Closure that constructs the office: $\lambda w\lambda t[^0King\_of_{wt}\ ^0France] \rightarrow \iota_{\tau\omega}$. But the property of being bald cannot be ascribed to an individual office. Instead it is ascribed to the individual (if any) occupying the office. Thus the office has to be subjected to intensional descent first: $\lambda w\lambda t[^0King\_of_{wt}\ ^0France]_{wt} \rightarrow_v \iota$. The property itself has to be extensionalized as well: $^0Bald_{wt}$. By Composing these two constructions, we obtain either a truth-value (**T** or **F**) or nothing, according as the King of France is, or is not, bald, or does not exist, respectively. Finally, by abstracting over the values of the variables w and t, we construct the proposition:

$$\lambda w\lambda t[^0Bald_{wt}\lambda w\lambda t[^0King\_of_{wt}\ ^0France]_{wt}]$$

This construction is assigned as its meaning to the Strawsonian variant of the sentence 'The King of France is bald'. So much for the basic notions of TIL and its method of analysis.

4 Definite Descriptions: Strawsonian or Russellian?

Now I am going to propose a solution to the Strawson-Russell standoff. In other words, I am going to analyse the phenomena of presupposition and entailment connected with using definite descriptions with supposition *de dicto* or *de re*, and I will show how the topic-focus distinction determines which of the two cases applies.

4.1 Topic-Focus Ambiguity

When used in a communicative act, an atomic sentence communicates something (the focus F) about something (the topic T). Thus the schematic structure of an atomic sentence is $F(T)$. The topic T of a sentence S is often associated with a presupposition P of S such that P is entailed both by S and $non - S$. On the other hand, the clause in the focus usually occasions a mere entailment of some P by S.[13]

To give an example, consider the sentence 'Our defeat was caused by John'. There are two possible readings of this sentence. Taken one way, the sentence is about our defeat, conveying the snippet of information that it was caused by John. In such a situation the sentence is associated with the presupposition that we were defeated. Indeed, the negated form of the sentence, 'Our defeat was not caused by John', also implies that we were defeated. Thus 'our defeat' is the topic and 'was caused by John' the focus clause. Taken the other way, the sentence is about the topic John,

[13] See Refs. [6, 7].

ascribing to him the property that he caused our defeat (focus). Now the scenario of truly asserting the negated sentence can be, for instance, the following. Though it is true that John has a reputation for being rather a bad player, Paul was in excellent shape and so we won. Or, another scenario is thinkable. We were defeated, only not because of John but because the whole team performed poorly. Hence, our being defeated is not presupposed by this reading, it is only entailed.

Schematically, if $\models$ is the relation of entailment, then the logical difference between a mere entailment and a presupposition is this:

P is a *presupposition* of S : $(S \models P)$ and $(non\text{-}S \models P)$

Thus if P is not true, then neither S nor *non-S* is true. Hence, S has no truth-value.

P is only *entailed* by S: $(S \models P)$ and neither $(non\text{-}S \models P)$ nor $(non\text{-}S \models non\text{-}P)$

Hence if S is not true we cannot deduce anything about the truth-value of P.

4.2 The King of France Revisited

Above we analyzed the sentence 'The King of France is bald' on its perhaps most natural reading as predicating the property of being bald (the focus) of the individual (if any) that is the present King of France (the topic). Yet there is another, albeit less natural reading of the sentence. Imagine that the sentence is uttered in a situation where we are talking about baldness, and somebody asks 'Who is bald?' The answer might be 'Well, among those who are bald there is the present King of France'. If you receive such an answer, you most probably protest, 'This cannot be true, for there is no King of France now'. On such a reading the sentence is about baldness (topic) claiming that this property is instantiated, among others, by the King of France (focus). Since there are no rigorous grammatical rules in English to distinguish between the two variants, the input of our *logical* analysis is the result of a *linguistic* analysis, where the topic and focus of a sentence are made explicit.[14] In this paper I mark the topic clause in italics. The two readings of the above sentence are:

(S) '*The king of France* is bald' (Strawsonian)
(R) 'The king of France is *bald*' (Russellian)

The analysis of (S) is as above: $\lambda w \lambda t[{}^0Bald_{wt} \lambda w \lambda t[{}^0King\_of_{wt}\ {}^0France]_{wt}]$.

The meaning of 'the King of France', viz. $\lambda w \lambda t[{}^0King\_of_{wt}\ {}^0France]$, occurs in (S) with *de re* supposition, because the object of predication is the unique *value* in the chosen $\langle w, t \rangle$-pair of evaluation of the office.[15] To construct this value

[14] For instance, in the Prague Dependency Treebank for the Czech language, the tectogrammatical representation contains the semantic structure of sentences with topic-focus annotators. For details, see http://ufal.mff.cuni.cz/pdt2.0/

[15] For details on the analysis of *de dicto* vs. *de re* supposition within TIL framework, see [5, esp. §§1.5.2 and 2.6.2] and also [3].

(if any), the office must be extensionalized. This is achieved in (S) by Composition $\lambda w\lambda t[^0King\_of_{wt}\ ^0France]_{wt}$.

The following *two de re principles* are satisfied: the principle of *existential presupposition* and the principle of *substitution of co-referential* expressions. Thus the following arguments are valid (though not sound):

$$\frac{\textit{The King of France} \text{ is (not) bald}}{\text{The King of France exists}}$$

$$\frac{\begin{array}{c}\textit{The King of France} \text{ is bald}\\ \textit{The King of France} \text{ is Louis XVI}\end{array}}{\textit{Louis XVI} \text{ is bald}}$$

To prove the validity of the first argument, we need to analyse its conclusion 'The King of France exists'. In TIL (non-trivial) existence is explicated as a property of intensions to be instantiated in a given $\langle w, t\rangle$-pair of evaluation.[16] Thus to say that unicorns do not exist is tantamount to saying that at the given world w and time t the property of being a unicorn has empty class of individuals as its extension. Similarly, that the King of France does not exist means that the office of the King of France is vacant at the world and time of evaluation.

Thus in our case we have $Exist/(o\iota_{\tau\omega})_{\tau\omega}$, the property of an office's being occupied at a given world/time pair that is defined as follows:

$$^0Exist =_{of} \lambda w\lambda t\lambda c[^0\exists\lambda x[x =_i c_{wt}]]$$

Types: $\exists/(o(o\iota))$; $c \rightarrow_v \iota_{\tau\omega}$; $x \rightarrow_v \iota$; $=_{of}\ /(o(o\iota_{\tau\omega})_{\tau\omega}(o\iota_{\tau\omega})_{\tau\omega})$: the identity of properties of individual offices; $=_i\ /(o\iota\iota)$: the identity of individuals, $x \rightarrow_v \iota$.

We introduce $Louis/\iota$, $Empty/(o(o\iota))$: the singleton containing the empty set of individuals, and $Improper/(o{*}_1)_{\tau\omega}$: the property of constructions of being v-improper at a given $\langle w, t\rangle$-pair; the other types are as above. Then for any $\langle w, t\rangle$-pair the following proof steps are truth-preserving:

(a) *existence:*

(1) $(\neg)[^0Bald_{wt}\lambda w\lambda t[^0King\_of_{wt}\ ^0France]_{wt}]$ $\varnothing$

(2) $\neg[^0Improper_{wt}{}^0[\lambda w\lambda t[^0King\_of_{wt}\ ^0France]_{wt}]]$ by Def. 2, iii)

(3) $\neg[^0Empty\ \lambda x[x =_i [\lambda w\lambda t[^0King\_of_{wt}\ ^0France]]_{wt}]]$ by Def. 2, iv)

(4) $[^0\exists\lambda x[x =_i [\lambda w\lambda t[^0King\_of_{wt}\ ^0France]]_{wt}]]$ EG

(5) $[^0Exist_{wt}[\lambda w\lambda t[^0King\_of_{wt}\ ^0France]]]$ by def. of *Exist*

[16] For details see [5], 2.3.

Remark Note that in step (2) the property of being *Improper* of type $(o*_1)_{\tau\omega}$ is applied to the *construction* $[\lambda w\lambda t[^0King\_of_{wt}\ ^0France]_{wt}]$ of type $*_1$ that is supplied here by its Trivialisation $^0[\lambda w\lambda t[^0King\_of_{wt}\ ^0France]_{wt}]$ belonging to type $*_2$. On the other hand in step (3) *Empty* of type $(o(o\iota))$ is applied to the set of individuals constructed here by $\lambda x\ [x =_i [\lambda w\lambda t[^0King\_of_{wt}\ ^0France]]_{wt}]$. These two steps are necessary in order to existentially generalize in step (4). In the logic of partial functions such as TIL we cannot carelessly generalize before proving that the set to which existential quantifier is applied is non-empty.

(b) *substitution:*

(1) $[^0Bald_{wt}\lambda w\lambda t[^0King\_of_{wt}\ ^0France]_{wt}]$ $\varnothing$
(2) $[^0Louis =_i \lambda w\lambda t[^0King\_of_{wt}\ ^0France]_{wt}]$ $\varnothing$
(3) $[^0Bald_{wt}\ ^0Louis]$ substitution of identicals

As explained above, the sentence (R) is not associated with the presupposition that the present King of France should exist, because 'the King of France' occurs now in the focus clause. The truth and falsity conditions of the Russellian 'The King of France is *bald*' are as follows:

- True, if and only if among those who are bald there is the King of France.
- False, if and only if among those who are bald there is no King of France (either because the King's office is not occupied, or its occupant is not bald).

Thus the two readings (S) and (R) have *different* truth-conditions, and they are not equivalent, albeit they are co-entailing. The reason is this. Trivially, by definition a valid argument is *truth-preserving from premises to conclusion*. However, due to partiality, the entailment relation may fail to be *falsity-preserving from conclusion to premises*. As a consequence, if A, B are constructions of propositions such that $A \models B$ and $B \models A$, then A, B are not necessarily equivalent in the sense of constructing the same proposition. Though the propositions take the truth-value **T** at exactly the same world/times, they may differ in such a way that at some $\langle w, t\rangle$-pair(s) one takes the value **F** while the other is undefined. The pair of meanings of (S) and (R) is an example of such co-entailing, yet non-equivalent hyperpropositions.

Next I am going to analyse (R). TIL makes it possible to avoid the other objections against Russell's analysis as well. The Russellian rephrasing of the sentence 'The King of France is bald' is this: 'There is a unique individual such that he is the King of France and he is *bald*'. This sentence expresses the construction[17]

(R*) $\lambda w\lambda t[^0\exists\lambda x[x =_i [\lambda w\lambda t[^0King\_of_{wt}\ ^0France]_{wt}] \wedge [^0Bald_{wt}x]]]$.

TIL analysis of the 'Russellian rephrasing' does not deprive 'the King of France' of its meaning. The meaning is invariably, in all contexts, the Closure $\lambda w\lambda t[^0King\_of_{wt}\ ^0France]$. Moreover, even the main objection that Russell simply

[17] Note that in TIL we do not need the construction corresponding to $\forall y(Fy \supset x = y)$ specifying the uniqueness of the King of France, because it is inherent in the meaning of 'the King of France'. The meaning of definite descriptions like 'the King of France' is a construction of an individual office of type $\iota_{\tau\omega}$ occupied in each $\langle w, t\rangle$-pair by at most one individual.

gets the truth-conditions wrong if there is no King of France is irrelevant, because in (R^*) the Closure $\lambda w \lambda t[{}^0 King\_of_{wt}\ {}^0 France]$ occurs intensionally (that is *de dicto*) unlike in the analysis of (S) where it occurs extensionally (*de re*).[18] The existential quantifier $\exists$ applies to *sets* of individuals rather than a particular individual. The proposition constructed by (R^*) is true if the *set* of individuals who are bald contains the individual who occupies the office of King of France, otherwise it is simply false. The truth conditions specified by (R^*) are Russellian. Thus we might be content with (R^*) as an adequate analysis of the Russellian reading (R). Yet we should not be. The reason is this. Russell's analysis has another defect; it does not comply with *Carnap's principle of subject-matter*, which states, roughly, that only those entities that receive mention in a sentence can become constituents of its meaning.[19] In other words, (R^*) is not the literal analysis of the sentence 'The King of France is *bald*', because existence and conjunction do not receive mention in the sentence. I am going to propose this literal analysis below. Yet before doing so, I must tackle still another issue in which Russell and Strawson differ, namely the problem of *negation*.

From a logical point of view, the two readings differ in the way their respective *negated* form is obtained. Whereas the Strawsonian negated form is 'The *King of France* is *not* bald', which obviously lacks a truth-value at those $\langle w, t \rangle$-pairs where the royal office is not occupied, the Russellian negated form is 'It is not true that the King of France is bald', which is true at those $\langle w, t \rangle$-pairs where the office is not occupied. Thus in the Strawsonian case the property of not being bald is ascribed to the individual, if any, that occupies the royal office. On the other hand, in the Russellian case the property of not being true is ascribed to the whole proposition that the King is bald, and thus (the same meaning of) the description 'the King of France' occurs with *de dicto* supposition. In order to ascribe the property of being true to the whole proposition, we apply the propositional property $True/(o o_{\tau\omega})_{\tau\omega}$ defined as follows: Let P be a propositional construction ($P/*_n \rightarrow o_{\tau\omega}$). Then $[{}^0 True_{wt} P]$ v-constructs **T** iff P_{wt} v-constructs **T**, otherwise **F**.[20] Now the analysis of the sentence (R) is this construction:

(R') $\lambda w \lambda t[{}^0 True_{wt} \lambda w \lambda t[{}^0 Bald_{wt} \lambda w \lambda t[{}^0 King\_of_{wt}\ {}^0 France]_{wt}]]$

Neither (R') nor its negation

(R'_neg) $\lambda w \lambda t \neg[{}^0 True_{wt} \lambda w \lambda t[{}^0 Bald_{wt} \lambda w \lambda t[{}^0 King\_of_{wt}\ {}^0 France]_{wt}]]$

entails that the King of France exists, which is just as it should be. (R'_neg) constructs the proposition *non-P* that takes the truth-value **T** if the proposition that the King

---

[18] For the definition of extensional, intensional and hyperintensional occurrence of a construction, see [5, §2.6].

[19] See [1, §24.2, §26] and [5, §2.1.1.].

[20] There are two other propositional properties of the same type, namely *False* and *Undefined*: $[{}^0 False_{wt} P]$ v-constructs the truth-value **T** iff $[\neg P_{wt}]$ v-constructs **T**, otherwise **F**. $[{}^0 Undef_{wt} P]$ v-constructs the truth-value **T** iff $[\neg[{}^0 True_{wt} P] \wedge \neg[{}^0 False_{wt} P]]$ v-constructs **T**, otherwise **F**.

of France is bald takes the value **F** (because the King of France is not bald) or is undefined (because the King of France does not exist)

To adduce a more natural example of topic/focus ambiguity, consider another sample sentence:

(2) '*The King of France* visited London yesterday.'

The topic phrase of (2) is 'the King of France'. Hence the sentence ascribes to the holder (if any) of the royal office at the world/time pair of evaluation the property of having visited London yesterday (the focus). Thus both (2) and its negation share the presupposition that the King of France actually exist *now* (that is, at the time of evaluation). If this presupposition fails to be satisfied, then neither of the propositions expressed by (2) and its negation '*The King of France* did not visit London yesterday' has a truth-value.

The situation is different in the case of the sentence (3):

(3) '*London* was visited by the King of France yesterday.'

Now the property (the focus) of having been visited by the King of France yesterday is predicated of London (the topic). The existence of the King of France at the time of evaluation is presupposed neither by (3) nor by its negation. The sentence can be read as 'Among the visitors of London yesterday was the then King of France'. The existence of the King of France *yesterday* is only entailed by (3) and not presupposed.[21] My analyses respect these conditions.

Let $Yesterday/((o\tau)\tau)$ be the function that associates a given time t with the time interval that is yesterday with respect to t; $Visit/(o\iota\iota)_{\tau\omega}$; $King\_of/(\iota\iota)_{\tau\omega}$; $France/\iota$; $\exists/(o(o\tau))$.

The analysis of (2) comes down to

(2*) $\lambda w\lambda t[\lambda x[{}^0\exists\lambda t^*[[[{}^0Yesterday\ t]t^*] \wedge [{}^0Visit_{wt^*}x\ {}^0London]]]$
$\lambda w\lambda t[{}^0King\_of_{wt}\ {}^0France]_{wt}]$

In (2*) the royal office is extensionalized with respect to the world w and the time t of evaluation. At such $\langle w, t\rangle$-pairs at which the office is not occupied the proposition constructed by (2*) has no truth-value, because the extensionalization of the office yields no individual, the Composition $\lambda w\lambda t[{}^0King\_of_{wt}\ {}^0France]_{wt}$ being v-improper. We have the Strawsonian case of the King's existence being presupposed. On the other hand, the sentence (3) expresses

(3*) $\lambda w\lambda t[{}^0\exists\lambda t^*[[[{}^0Yesterday\ t]t^*] \wedge [{}^0Visit_{wt^*}$
$\lambda w\lambda t[{}^0King\_of_{wt}\ {}^0France]_{wt^*}\ {}^0London]]]$

In (3*) the royal office is extensionalized with respect to world w and time t^* belonging to the interval $[{}^0Yesterday\ t]$. If the office goes vacant for all such t^* the Composition $\lambda w\lambda t[{}^0King\_of_{wt}\ {}^0France]_{wt^*}$ is v-improper for any t^* belonging to

[21] [21] disregards this reading, saying that any sentence containing 'the King of France' comes with the presupposition that the King of France exist *now*. In my opinion, this is because he considers only the *neutral* reading, thus disregarding topic-focus ambiguities.

$[^0Yesterday\ t]$. Hence the time interval v-constructed by the Closure $\lambda t^*[[[^0Yesterday\ t]t^*] \wedge [^0Visit_{wt^*} \lambda w \lambda t[^0King\_of_{wt}\ ^0France]_{wt^*}\ ^0London]]$ is empty and the existential quantifier takes this interval to **F**. On the other hand, at such a $\langle w, t \rangle$-pair at which the proposition constructed by (3*) is true, the Composition $[^0\exists \lambda t^*[[[^0Yesterday\ t]t^*] \wedge [^0Visit_{wt^*} \lambda w \lambda t[^0King\_of_{wt}\ ^0France]_{wt^*}\ ^0London]]]$ v-constructs **T**. This means that the second conjunct v-constructs **T** as well and the Composition $\lambda w \lambda t[^0King\_of_{wt}\ ^0France]_{wt^*}$ is not v-improper. Thus the royal office is occupied *at some time* t^* belonging to $[^0Yesterday\ t]$. This is as it should be, because (3*) *only entails the existence* of the King of France *yesterday*. We have the Russellian case: the meaning of 'the King of France' occurs with *de dicto* supposition with respect to the temporal parameter t.

5 Conclusion

In this paper I demonstrated that both the proponents of Russell's quantificational analysis and of Strawson's referential analysis of definite descriptions are partly right and partly wrong, because sentences of the form 'The *F* is a *G*' are systematically ambiguous. Their ambivalence stems from different topic-focus articulation, and I brought out the *semantic*, as opposed to pragmatic, character of this ambivalence. I showed that a definite description occurring in the topic of a sentence with *de re* supposition corresponds to the Strawsonian analysis of definite descriptions, while a definite description occurring in the focus with *de dicto* supposition corresponds to the Russellian analysis. While the clause standing in topic position triggers a presupposition, a focus clause usually only entails rather than presupposes another proposition. The procedural semantics of TIL provides rigorous analyses such that sentences differing only in their topic-focus articulation are assigned different constructions producing different propositions (truth-conditions) and having different consequences.

Moreover, the proposed analysis of the Russellian reading does not deprive definite descriptions of their meaning. Just the opposite; 'the *F*' receives a context-invariant meaning, which is the construction of an individual office. What is dependent on context is the way this (one and the same) construction is used. Thus I also demonstrated that Donnellan-style referential and attributive uses of an occurrence of 'the F' do not bring about a shift of meaning of 'the F'. Instead, one and the same context-invariant meaning is a constituent of different procedures that behave in logically different ways.

Acknowledgments This research was funded by the internal grant agency of VSB-TU of Ostrava, project No. SP2014/157, 'Knowledge modelling, process simulation and design'. The present paper is a revised and improved version of a part of the book chapter, see http://www.intechopen.com/books/semantics-in-action-applications-and-scenarios/resolving-ambiguities-in-natural-language. I am grateful to an anonymous reviewer for valuable comments that improved to quality of the paper.

References

1. Carnap, R. (1947). *Meaning and necessity*. NULL: The University of Chicago Press.
2. Donnellan, K. S. (1966). Reference and definite descriptions. *Philosophical Review*, *75*(3), 281–304.
3. Duží, M. (2004). Intensional logic and the irreducible contrast between de dicto and de re. *ProFil* *5*(1), 1–34. ISSN 1212–9097. http://profil.muni.cz/01_2004/duzi_de_dicto_de_re.pdf
4. Duží, M., Jespersen, B., & Materna, P. (2010a). The logos of semantic structure. In Stalmaszczyk, P. (Ed.) *Philosophy of language and linguistics* (Vol. 1, pp. 85–102). Frankfurt: The Formal Turn, Ontos Verlag.
5. Duží, M., Jespersen, B., & Materna, P. (2010b). *Procedural semantics for hyperintensional logic*. Berlin: Springer.
6. Gundel, J. K. (1999). Topic, focus and the grammar pragmatic interface. In J. Alexander, N. Han & M. Minnick (Eds.), *Proceedings of the 23rd Annual Penn Linguistics Colloquium* (Vol. 6.1, pp. 185–200). Penn Working Papers in Linguistics.
7. Hajičová, E. (2008). What we are talking about and what we are saying about it. In A. Gelbukh (Ed.), *Computational linguistics and intelligent text processing* (pp. 241–262). Berlin, Heidelberg: Springer.
8. Jespersen, B. (2008). Predication and extensionalization. *Journal of Philosophical Logic*, *37*(5), 479–499.
9. Jespersen, B. (2010). How hyper are hyperpropositions? *Language and Linguistics Compass*, *4*(2), 96–106.
10. Jespersen, B. (2012). Recent work on structured meaning and propositional unity. *Philosophy Compass*, *7*(9), 620–630.
11. Kripke, S. A. (1977). Speaker's reference and semantic reference. In P. French, T. E. Uehling & H. K. Wettstein (Eds.), *Contemporary perspectives in the philosophy of language* (pp. 6–27). Minneapolis: University of Minnesoty Press.
12. Ludlow, P. (2007). Descriptions. In E. N. Zalta (Ed.), *The Stanford encyclopedia of philosophy*. http://plato.stanford.edu/entries/descriptions
13. Neale, S. (1990). *Descriptions*. Cambridge, MA: MIT Press.
14. Russell, B. (1905). On denoting. *Mind*, *14*(4), 479–493.
15. Russell, B. (1957). Mr. Strawson on referring. *Mind*, *66*(263), 385–389.
16. Strawson, P. F. (1950). On referring. *Mind*, *59*(235), 320–334.
17. Strawson, P. F. (1952). *Introduction to logical theory*. London: Routledge.
18. Strawson, P. F. (1964). Identifying reference and truth-values. *Theoria*, *30*(2), 96–118.
19. Tichý, P. (1988). *The foundations of Frege's logic*. Walter de Gruyter.
20. Tichý, P. (2004). *Collected papers in logic and philosophy*. In V. Svoboda, B. Jespersen & C. Cheyne (eds.), Prague: Filosofia, Czech Academy of Sciences, and Dunedin: University of Otago Press.
21. von Fintel, K. (2004). Would you believe it? The king of France is back! (presuppositions and truth-value intutions). In M. Reimer & A. Bezuidenhout (Eds.), *Descriptions and beyond* (pp. 315–341). Oxford: Clarendon Press.

Tableau Metatheorem for Modal Logics

Tomasz Jarmużek

Abstract The aim of the paper is to demonstrate and prove a tableau metatheorem for modal logics. While being effective tableau methods are usually presented in a rather intuitive way and our ambition was to expose the method as rigorously as possible. To this end all notions displayed in the sequel are couched in a set theoretical framework, for example: branches are sequences of sets and tableaus are sets of these sequences. Other notions are also defined in a similar, formal way: maximal, open and closed branches, open and closed tableaus. One of the distinctive features of the paper is introduction of what seems to be the novelty in the literature: the notion of tableau consequence relation. Thanks to the precision of tableau metatheory we can prove the following theorem: completeness and soundness of tableau methods are immediate consequences of some conditions put upon a class of models **M** and a set of tableau rules **MRT**. These conditions will be described and explained in the sequel. The approach presented in the paper is very general and may be applied to other systems of logic as long as tableau rules are defined in the style proposed by the author. In this paper tableau tools are treated as an entirely syntactical method of checking correctness of arguments [1, 2].

Keywords Modal logics · Possible world's semantics · Tableau rules · Branch · Open branch · Closed branch · Maximal branch · Open tableau · Closed tableau · Tableau metatheorem

1 Basic Notions

In this part of the paper we remind some standard semantic notions and we introduce some new ones that will be necessary to formulate and prove facts about tableaus. The traditional focus of modal logic has been completeness with respect to classes

T. Jarmużek (✉)
Department of Logic, Nicolaus Copernicus University, Toruń, Poland
e-mail: jarmuzek@umk.pl

R. Ciuni et al. (eds.), *Recent Trends in Philosophical Logic*,
Trends in Logic 41, DOI: 10.1007/978-3-319-06080-4_8,

of structures. Somewhat differently, we concentrate on completeness with respect to classes of models, since from the point of view of our approach it is more convenient to carry out proofs and define notions for classes of models rather than of structures. However, it should be emphasized that the former, more traditional approach can be easily translated into ours.[1]

It was our intention not to include in the paper any decidability issues whatsoever. This topic of utmost importance for tableau methods theory is simply too complex to be developed in one paper. Nevertheless, our opinion is that tools we define can be useful in treating decidability problems, since with formal definitions of the key notions at hand in infinite cases we are able to define cycles of branches—sequences of applications of tableau rules that result in infinite branches. For this reason, we find a formal theory of tableaus as a necessary condition of precise approach to the problem of tableau decidability.

1.1 Semantics

Let For be the set of all modal formulas build over the alphabet: $\mathsf{Var} \cup \{\neg, \wedge, \vee, \rightarrow, \leftrightarrow, \Diamond, \Box\}$. Let $\mathfrak{M} = \langle W, R, V, w\rangle$ be a possible world model.[2] Here we have a standard definition of being true in a model.

Definition 1.1 (*Truth in model*) Let $\mathfrak{M} = \langle W, R, V, w\rangle$ be a model and $A \in \mathsf{For}$. We say that A is true in $\mathfrak{M}$ (in short: $\mathfrak{M} \models A$) iff for all $B, C \in \mathsf{For}$

1. if $A \in \mathsf{Var}$, then $V(A, w) = 1$
2. if $A := \neg B$, then B is not true in $\mathfrak{M}$ (in short: $\mathfrak{M} \not\models B$)
3. if $A := (B \wedge C)$, then $\mathfrak{M} \models B$ and $\mathfrak{M} \models C$
4. if $A := (B \vee C)$, then $\mathfrak{M} \models B$ or $\mathfrak{M} \models C$
5. if $A := (B \rightarrow C)$, then $\mathfrak{M} \not\models B$ or $\mathfrak{M} \models C$
6. if $A := (B \leftrightarrow C)$, then $\mathfrak{M} \models B$ iff $\mathfrak{M} \models C$
7. if $A := \Box B$, then $\forall_{u \in W}(wRu \Longrightarrow \langle W, R, V, u\rangle \models B)$
8. if $A := \Diamond B$, then $\exists_{u \in W}(wRu \;\&\; \langle W, R, V, u\rangle \models B)$.

Let X be a set of formulas and $\mathfrak{M}$ be a model. We say that X is *true in* $\mathfrak{M}$ (in short: $\mathfrak{M} \models X$) iff for all $A \in X$, $\mathfrak{M} \models A$. We say that a set of formulas is *inconsistent* iff for any model $\mathfrak{M} \not\models X$. Otherwise, we call X *consistent*.[3]

Fact 1.2 *For any formula A and any set of formulas X, if* $\{A, \neg A\} \subseteq X$*, then X is inconsistent.*

---

[1] The author would like to thank an anonymous reviewer for many valuable comments and suggestions that have allowed to improve the paper.

[2] In the literature this kind of model is usually called *a pointed model*, but we will shortly call it model.

[3] We use a word *inconsistent* instead of—for example—*contradictory*, since it enables us to do a direct transition between semantic and tableau notions.

We define a consequence relation $\models_{\mathbf{M}}$ on $2^{\mathsf{For}} \times \mathsf{For}$, where $\mathbf{M}$ is a class of models.

Definition 1.3 Let $A \in \mathsf{For}$ and $X \subseteq \mathsf{For}$. Let $\mathbf{M}$ be a class of models. We say that A is a consequence of X modulo $\mathbf{M}$ (in short: $X \models_{\mathbf{M}} A$) iff $\forall_{\mathfrak{M} \in \mathbf{M}}(\mathfrak{M} \models X \Longrightarrow \mathfrak{M} \models A)$.

1.2 Tableau Rules

In order to define precise notions of tableau proofs we need some auxiliary notions. First of all, we define a language of tableau proofs, the set of expressions.

Definition 1.4 (*Expressions*) The set of expressions Ex is the smallest set that includes all elements of:

- Cartesian product: $\mathsf{For} \times \mathbb{N}$
- $\{irj : i, j \in \mathbb{N}\}$
- $\{\sim irj : i, j \in \mathbb{N}\}$
- $\{i = j : i, j \in \mathbb{N}\}$
- $\{\sim i = j : i, j \in \mathbb{N}\}$

where $\mathbb{N}$ is the set of natural numbers. The elements of $\mathbb{N}$ we call *indexes*.

To define modal tableau rules we need a definition of similar sets of expressions. Firstly, we define some function choosing indexes from the set of expressions Ex.

Definition 1.5 (*Function choosing indexes*) The function choosing indexes we call a function $* : P(\mathsf{Ex}) \longrightarrow P(\mathbb{N})$ defined for any $X \subseteq \mathsf{Ex}$, $A \in \mathsf{For}$ and $i, j \in \mathbb{N}$ by conditions:

- $*(\emptyset) = \emptyset$
- $*(\{\langle A, i\rangle\}) = \{i\}$
- $*(\{irj\}) = \{i, j\}$
- $*(\{\sim irj\}) = \{i, j\}$
- $*(\{i = j\}) = \{i, j\}$
- $*(\{\sim i = j\}) = \{i, j\}$
- $*(X) = \bigcup\{*(\{x\}) : x \in X\}$, if $|X| > 1$

For any subset of Ex function $*$ collects all indexes occurring in expressions in this set.
We now introduce a notion of similar sets of expressions. Shortly speaking, two sets of expressions are similar iff exactly the same formulas occur in their expressions and all expressions in both sets are structurally similar on indexes.

Definition 1.6 (*Similar sets of expressions*) For any two sets of expressions X, Y, we say that X is *similar to* Y iff there is a bijection $g : *(X) \longrightarrow *(Y)$ (where $*(X)$, $*(Y)$ are the sets of indexes occurring in expressions of X and Y) such that for any $A \in \mathsf{For}$, $i, j \in \mathbb{N}$:

- $\langle A, i\rangle \in X$ iff $\langle A, g(i)\rangle \in Y$
- $irj \in X$ iff $g(i)rg(j) \in Y$
- $\sim irj \in X$ iff $\sim g(i)rg(j) \in Y$
- $i = j \in X$ iff $g(i) = g(j) \in Y$
- $\sim i = j \in X$ iff $\sim g(i) = g(j) \in Y$.

Corollary 1.7 *The relation of being similar defined on sets of expressions is an equivalence relation, i.e. reflexive, transitive, and symmetrical.*

Definition 1.8 (*Tableau inconsistent sets of expressions*) Let $X \subseteq \mathsf{Ex}$. We say that X is tableau inconsistent iff for some $A \in \mathsf{For}$, $i, j \in \mathbb{N}$ one of the conditions is fulfilled:

1. $\langle A, i\rangle, \langle \neg A, i\rangle \in X$
2. $irj, \sim irj \in X$
3. $i = j, \sim i = j \in X$.

Otherwise, we call set X tableau consistent. We shortly say that X is t-consistent or respectively t-inconsistent.

Corollary 1.9 *For any two sets of expressions X, Y, if X is similar to Y, then X is t-inconsistent iff Y is t-inconsistent.*

Moreover, we require some notion that connects models with sets of expressions.

Definition 1.10 (*Model satisfying a set of expressions*) Let $\mathfrak{M} = \langle W, R, V, w\rangle$ and $X \subseteq \mathsf{Ex}$. We say that $\mathfrak{M}$ *satisfies* X iff there is a function $f : \mathbb{N} \longrightarrow W$ such that for any $A \in \mathsf{For}$, $i, j \in \mathbb{N}$:

- if $\langle A, i\rangle \in X$, then $\langle W, R, V, f(i)\rangle \models A$
- if $irj \in X$, then $f(i)Rf(j)$
- if $\sim irj \in X$, then it is not that $f(i)Rf(j)$
- if $i = j \in X$, then $f(i)$ is equal to $f(j)$
- if $\sim i = j \in X$, then $f(i)$ is different from $f(j)$.

Fact 1.11 *Let X be a tableau inconsistent set of expressions. Then there is no model $\mathfrak{M}$ satisfying X.*

Proof By definitions of tableau inconsistent set of expressions 1.8, model satisfying set of expressions 1.10 and definition of truth in model 1.1. □

Now, we can give a very general notion of a modal tableau rule. However, we first comment some example of the rule:

$R_\wedge$: $\dfrac{X \cup \{\langle (A \wedge B), i\rangle\}}{X \cup \{\langle (A \wedge B), i\rangle, \langle A, i\rangle, \langle B, i\rangle\}}$, where $X \cup \{\langle (A \wedge B), i\rangle\}$ is a t-consistent

set of expressions and $\{\langle A, i\rangle, \langle B, i\rangle\} \not\subseteq X$.

The rule can be applied only to t-consistent sets of expressions, the output set is a proper superset of the initial set, and the rule captures all possible sets that are instances of this schema. The additional properties of rules are expressed in the following general definition.

Definition 1.12 (*Tableau rule*) Let $P(\mathsf{Ex})$ be a power set of Ex. Let $P(\mathsf{Ex})^n$ be an n–ary Cartesian product $\underbrace{P(\mathsf{Ex}) \times \cdots \times P(\mathsf{Ex})}_{n}$, for some $n \in \mathbb{N}$.

- A *tableau rule* is any subset $R \subseteq P(\mathsf{Ex})^n$, for some $n \geq 2$, such that if $\langle X_1, \ldots, X_n \rangle \in R$, then:
  - $X_1 \subset X_i$, for all $1 < i \leq n$
  - X_1 is t-consistent[4]
  - if $k \neq l$, then $X_k \neq X_l$, for all $1 < k, l \leq n$
  - (Closure under Similarity) for all sets of expressions Y_1 such that Y_1 is similar to X_1:
    1. there are sets of expressions Y_2, ..., Y_n, such that $\langle Y_1, \ldots, Y_n \rangle \in R$
    2. and for all $1 < i \leq n$, Y_i is similar to X_i
  - (Existence of Core) for some finite $Y \subseteq X_1$
    1. there exists exactly one such n-tuple $\langle Z_1, \ldots, Z_n \rangle \in R$ that $Z_1 = Y$
    2. there is no proper subset $U_1 \subset Y$ and n-tuple $\langle U_1, \ldots, U_n \rangle \in R$
    3. for any $1 < i \leq n$, $Z_i = Z_1 \cup (X_i \setminus X_1)$
  - (Closure under Expansion) for any t-consistent set of expressions Z_1 such that $X_1 \subset Z_1$ and for all $1 < i \leq n$, X_i is not similar to any subset of Z_1:
    if for some finite $Y \subseteq X_1$:
    1. there exists exactly one such n-tuple $\langle W_1, \ldots, W_n \rangle \in R$ that $W_1 = Y$
    2. there is no proper subset $U_1 \subset Y$ and n-tuple $\langle U_1, \ldots, U_n \rangle \in R$
    3. for any $1 < i \leq n$, $W_i = W_1 \cup (X_i \setminus X_1)$

    then:
    1. there are exactly $n - 1$ such sets of expressions Z_2, ..., Z_n that $\langle Z_1, \ldots, Z_n \rangle \in R$
    2. and for all $1 < i \leq n$, W_i is similar to $W_1 \cup (Z_i \setminus Z_1)$
  - (Closure under Finite Sets) if X_1 is a finite set, then for all $1 < i \leq n$, X_i is a finite set
  - (Closure under Finite Subsets) if X_1 is an infinite set, then
    1. there are finite sets of expressions Y_1, ..., Y_n, such that $\langle Y_1, \ldots, Y_n \rangle \in R$
    2. $Y_1 \subset X_1$
    3. and for all $1 < i \leq n$, $Y_i = Y_1 \cup (X_i \setminus X_1)$.
- By saying that a rule R was applied to X_1, we mean that for some $1 < i \leq n$, exactly one X_i of $\langle X_1, \ldots, X_n \rangle$ was chosen.

Now, we need a definition of a proper part of a rule, called a core of rule.

[4] When we impose the condition of t-consistency, it seems we are not able to capture the sound rule $\{A, \neg A\} \models B$. This is not the case, because—as we will see—starting from a set of tableau premisses $\{\langle A, i\rangle, \langle \neg A, i\rangle, \langle \neg B, i\rangle\}$, for some formulas $A, B \in \mathsf{For}$ and some index $i \in \mathbb{N}$, immediately we have a closed tableau, so the rule is sound.

Definition 1.13 (*Core of rule*) Let $R \in \mathbf{MRT}$ and $\langle X_1, \ldots, X_n \rangle \in R$, for some $n \in \mathbb{N}$. We say that $\langle Z_1, \ldots, Z_n \rangle \in R$ is *a core of the rule R in the set* $\langle X_1, \ldots, X_n \rangle$ iff

1. $Z_1 \subseteq X_1$
2. there is no proper subset $U_1 \subset Z_1$ and n-tuple $\langle U_1, \ldots, U_n \rangle \in R$
3. for any $1 < i \leq n$, $Z_i = Z_1 \cup (X_i \setminus X_1)$.

By definition of tableau rules 1.12 (Existence of Core) we have a corollary.

Corollary 1.14 *Let $R \in$ **MRT** and $\langle X_1, \ldots, X_n \rangle \in R$, for some $n \in \mathbb{N}$. There exists exactly one n-tuple $\langle Y_1, \ldots, Y_n \rangle$ that is a core of the rule R in the set $\langle X_1, \ldots, X_n \rangle$.*

Now, we define some more technical terminology. Let $X \subseteq \mathsf{Ex}$ be a set of expressions and $\mathbf{R}$ be a set of tableau rules. By $\mathbf{R}_X$ we denote a set of all rules in $\mathbf{R}$ that can be applied to X. Formally, $R \in \mathbf{R}_X$ iff $R \in \mathbf{R}$ and there is some n-tuple $\langle Y_1, \ldots, Y_n \rangle \in R$, such that $Y_1 = X$.

Let $R \in \mathbf{R}_X$, by R_X we denote a set of all n-tuples in R, such that their first member is X and if other members of two n-tuples in R_X differ, than the rule has two different cores. Formally, for any $n \in \mathbb{N}$, $\langle Y_1, \ldots, Y_n \rangle \in R_X$ iff:

- $\langle Y_1, \ldots, Y_n \rangle \in R$ and $Y_1 = X$
- for any Z_1, ..., $Z_n \subseteq \mathsf{Ex}$, if:
  - $\langle Z_1, \ldots, Z_n \rangle \in R_X$
  - $\langle Y_1, \ldots, Y_n \rangle \neq \langle Z_1, \ldots, Z_n \rangle$
  - $\langle Y'_1, \ldots, Y'_2 \rangle$ is a core of R in $\langle Y_1, \ldots, Y_n \rangle$
  - $\langle Z'_1, \ldots, Z'_n \rangle$ is a core of R in $\langle Z_1, \ldots, Z_n \rangle$

then $Y'_1 \neq Z'_1$.
Now, we can define a notion of modal tableau rules.

Definition 1.15 (*Modal Tableau Rules*) Let $\mathbf{R}$ be a set of tableau rules. We say that $\mathbf{R}$ is a set of modal tableau rules (in short: **MRT**) iff

1. $\mathbf{R}$ is finite[5]
2. for any set $X \subseteq \mathsf{Ex}$, if X is finite, then for any $R \in \mathbf{R}_X$, R_X is finite.

1.3 Examples of Rules

Here, we shall give some examples of **MRT** rules. They are well-known, but written in new forms. Let $X \subseteq \mathsf{Ex}$, $A, B \in \mathsf{For}$, $i, j \in \mathbb{N}$. Examples of **MRT** rules are rules defined by these schemas, where initial sets are *t-consistent*.

[5] It does not mean that the set of all instances of any rule in $\mathbf{R}$ is finite.

1.3.1 Classical Rules

$$R_{\wedge}: \frac{X \cup \{\langle (A \wedge B), i\rangle\}}{X \cup \{\langle (A \wedge B), i\rangle, \langle A, i\rangle, \langle B, i\rangle\}}$$

$$R_{\vee}: \frac{X \cup \{\langle (A \vee B), i\rangle\}}{X \cup \{\langle (A \vee B), i\rangle, \langle A, i\rangle\} || X \cup \{\langle (A \vee B), i\rangle, \langle B, i\rangle\}}$$

$$R_{\rightarrow}: \frac{X \cup \{\langle (A \rightarrow B), i\rangle\}}{X \cup \{\langle (A \rightarrow B), i\rangle, \langle \neg A, i\rangle\} || X \cup \{\langle (A \rightarrow B), i\rangle, \langle B, i\rangle\}}$$

$$R_{\leftrightarrow}: \frac{X \cup \{\langle (A \leftrightarrow B), i\rangle\}}{X \cup \{\langle (A \leftrightarrow B), i\rangle, \langle A, i\rangle, \langle B, i\rangle\} || X \cup \{\langle (A \leftrightarrow B), i\rangle, \langle \neg A, i\rangle, \langle \neg B, i\rangle\}}$$

$$R_{\neg\neg}: \frac{X \cup \{\langle \neg\neg A, i\rangle\}}{X \cup \{\langle \neg\neg A, i\rangle, \langle A, i\rangle\}}$$

$$R_{\neg\wedge}: \frac{X \cup \{\langle \neg(A \wedge B), i\rangle\}}{X \cup \{\langle \neg(A \wedge B), i\rangle, \langle \neg A, i\rangle\} || X \cup \{\langle \neg(A \wedge B), i\rangle, \langle \neg B, i\rangle\}}$$

$$R_{\neg\vee}: \frac{X \cup \{\langle \neg(A \vee B), i\rangle\}}{X \cup \{\langle \neg(A \vee B), i\rangle, \langle \neg A, i\rangle, \langle \neg B, i\rangle\}}$$

$$R_{\neg\rightarrow}: \frac{X \cup \{\langle \neg(A \rightarrow B), i\rangle\}}{X \cup \{\langle \neg(A \rightarrow B), i\rangle, \langle A, i\rangle, \langle \neg B, i\rangle\}}$$

$$R_{\neg\leftrightarrow}: \frac{X \cup \{\langle \neg(A \leftrightarrow B), i\rangle\}}{X \cup \{\langle \neg(A \leftrightarrow B), i\rangle, \langle \neg A, i\rangle, \langle B, i\rangle\} || X \cup \{\langle \neg(A \leftrightarrow B), i\rangle, \langle A, i\rangle, \langle \neg B, i\rangle\}}$$

It may seem that our approach is similar to the approach offered by Hintikka (so called Hintikka's sets). However, when our rules are applied then all previous elements are gathered together. Within Hintikka's approach former elements are finally abandoned. Within ours one goes through from one set to its extension.

1.3.2 Modal Rules

$$R_{\neg\Box}: \frac{X \cup \{\langle \neg\Box A, i\rangle\}}{X \cup \{\langle \neg\Box A, i\rangle, \langle \Diamond\neg A, i\rangle\}}$$

$$R_{\neg\Diamond}: \frac{X \cup \{\langle \neg\Diamond A, i\rangle\}}{X \cup \{\langle \neg\Diamond A, i\rangle, \langle \Box\neg A, i\rangle\}}$$

$$R_{\Box}: \frac{X \cup \{\langle \Box A, i\rangle, irj\}}{X \cup \{\langle \Box A, i\rangle, irj, \langle A, j\rangle\}}$$

The first variant of a rule for $\Diamond$:

$$R1_{\Diamond}: \frac{X \cup \{\langle \Diamond A, i\rangle\}}{X \cup \{\langle \Diamond A, i\rangle, irj, \langle A, j\rangle\}}, \text{ where:}$$

1. $j \notin *(X \cup \{\langle \Diamond A, i\rangle\})$
2. for any $k \in \mathbb{N}$, $\{irk, \langle A, k\rangle\} \nsubseteq X$

Within the range of Definition 1.12 we can define other modal tableau rules to express different properties of the accessibility relation. For example we may want to have a rule for models with an empty relation:
The second variant for $\Diamond$:

$$R2_{\Diamond}: \frac{X \cup \{\langle \Diamond A, i \rangle\}}{X \cup \{\langle \Diamond A, i \rangle, \langle \neg \Diamond A, i \rangle\}}$$

There are many other examples of modal tableau rules that define properties of relation R in a model. Below we have few examples. Let $X \subseteq \mathsf{Ex}$, $A \in \mathsf{For}$, $i, j, k \in \mathbb{N}$:

(Symmetry)

$$\frac{X \cup \{irj\}}{X \cup \{irj, jri\}} \qquad \forall_{w_1, w_2 \in W}(w_1 R w_2 \Rightarrow w_2 R w_1)$$

(Transitivity):

$$\frac{X \cup \{irj, jrk\}}{X \cup \{irj, jrk, irk\}} \qquad \forall_{w_1, w_2, w_3 \in W}(w_1 R w_2 \,\&\, w_2 R w_3 \Rightarrow w_1 R w_3)$$

(Reflexivity):

$$\frac{X \cup \{\langle A, i \rangle\}}{X \cup \{\langle A, i \rangle, iri\}} \qquad \forall_{w_1 \in W}(w_1 R w_1)$$

(Irreflexivity):

$$\frac{X \cup \{iri\}}{X \cup \{iri, \langle B, i \rangle, \langle \neg B, i \rangle, \}}, \text{ for some } B \in \mathsf{For} \qquad \forall_{w_1 \in W}(\sim w_1 R w_1)$$

(Antisymmetry):

$$\frac{X \cup \{irj, jri\}}{X \cup \{irj, jri, \langle B, i \rangle, \langle \neg B, i \rangle\}}, \text{ for some } B \in \mathsf{For} \ \forall_{w_1, w_2 \in W}(w_1 R w_2 \Rightarrow \sim w_2 R w_1)$$

There are many ways of defining some set of **MRT**. Its content depends only on our decisions and intentions, but all members of each **MRT** should satisfy the conditions of general Definitions 1.12 and 1.15.

1.4 Modal Branches and Tableaus

Given a set **MRT** we are able to define precisely notions of: branch, maximal branch, open branch, closed branch, tableau, complete tableau, closed tableau, and finally tableau consequence $\rhd_{\mathbf{MRT}}$. Each of those notions is determined by some set **MRT**.

A branch is a sequence of extending sets of expressions.

Definition 1.16 (*Branch*) • Let $K = \mathbb{N}$ or $K = \{1, 2, \ldots, n\}$, for some $n \in \mathbb{N}$. Let $X \subseteq \mathsf{Ex}$ and **MRT** be some set of modal tableau rules. A branch (a branch starting from X) is any sequence $\phi : K \longrightarrow P(\mathsf{Ex})$ that satisfies conditions:

- $\phi(1) = X$
- for any $i \in K$, if $i+1 \in K$, then there is a rule $R \in$ **MRT** and $\langle Y_1, \ldots, Y_m \rangle \in R$ such that $\phi(i) = Y_1$ and $\phi(i+1) = Y_k$, for some $m \in \mathbb{N}$ and some $1 < k \leq m$.

• Branches will be denoted by small Greek letters: ϕ, ψ, while sets of branches by big Greek letters: Φ, Ψ etc.
• Writing, for example, ϕ_K, ψ_M we will inform that a branch ϕ has a domain K, a branch ψ has a domain M etc.

We see that any branch is always modulo some set **MRT**. Therefore, writing about branches and more complex objects we should underline that fact. In further definitions we sometimes omit **MRT**, since those definitions are general, but in fact we always have **MRT**-branches, **MRT**-branch consequence relation, **MRT**-tableaus etc., always for some fixed set **MRT**. We can observe that any $X \subseteq \mathsf{Ex}$ is a one-member-branch, by Definition of branch 1.16.

Definition 1.17 (*Addition of branches*) Let $\phi : \{1, \ldots, n\} \longrightarrow P(\mathsf{Ex})$ and $\psi : \{1, \ldots, m\} \longrightarrow P(\mathsf{Ex})$ be branches, for some $n, m \in \mathbb{N}$, and let $\phi(n) = \psi(1)$. By $\phi \oplus \psi$ we mean a function $\varphi : \{1, \ldots, n, n+1, \ldots, n+m-1\} \longrightarrow P(\mathsf{Ex})$ defined as follows:

1. for any $1 \leq i \leq n$, $\varphi(i) = \phi(i)$
2. for any $n+1 \leq i \leq n+m-1$, $\varphi(i) = \psi((i-n)+1)$.

Corollary 1.18 *Let $\phi : \{1, \ldots, n\} \longrightarrow P(\mathsf{Ex})$ and $\psi : \{1, \ldots, m\} \longrightarrow P(\mathsf{Ex})$ be branches, for some $n, m \in \mathbb{N}$, and let $\phi(n) = \psi(1)$. Then $\phi \oplus \psi$ is a branch.*

Proof By Definition of branch 1.16. □

A closed/open branch is a branch that has got a t-inconsistent member/all t-consistent members.

Definition 1.19 (*Closed/open branch*) Let $\phi : K \longrightarrow P(\mathsf{Ex})$ be a branch. We say that ϕ is closed iff $\phi(n)$ is a t-inconsistent set for some $n \in K$. A branch is open iff is not closed.

Fact 1.20 *Let $\phi : K \longrightarrow P(\mathsf{Ex})$ be a closed branch. Then $K = \{1, 2, 3, \ldots, n\}$, for some $n \in \mathbb{N}$ (the branch is finite).*

Proof We take any closed branch $\phi : K \longrightarrow P(\mathsf{Ex})$. Since for some $n \in K$, $\phi(n)$ is a t-inconsistent set of expressions, so no rule of **MRT**, by 1.12, can make ϕ longer. In consequence, $n \in \mathbb{N}$ is the last member of ϕ and ϕ is finite. □

A maximal branch—intuitively, a branch that can not be longer. However, we have two variants, that we shall discuss. Formally, the first definition says:

Definition 1.21 (*Maximal branch—an initial definition*) Let $\phi : K \longrightarrow P(\mathsf{Ex})$ be a branch. We say that ϕ is maximal iff

1. $K = \{1, 2, 3, \ldots, n\}$, for some $n \in \mathbb{N}$
2. there is no branch ψ such that $\phi \subset \psi$.

This definition says that a maximal branch is finite and there is not a super branch. It is good for those logics in which from a finite set of expressions applying tableau rules always gives finite branches. Hence, it is not good for modal logics. For example, applying to the formula $\neg(\Diamond p \rightarrow \Diamond\Box p)$ rules for the boolean connectives, modalities and transitivity we never end a branch, but it can be maximal in other sense.

Example 1.22 We take a formula $\neg(\Diamond p \rightarrow \Diamond\Box p)$. We shall show that a branch obtained by use of rules for modal, boolean connectives and the rule for transitivity does not end.

1. $\langle\neg(\Diamond p \rightarrow \Diamond\Box p), 0\rangle$

$\downarrow$

2. $\langle\Diamond p, 0\rangle$, $\langle\neg\Diamond\Box p, 0\rangle$ — 1., $R_{\neg\rightarrow}$

$\downarrow$

3. $\langle\Box\neg\Box p, 0\rangle$ — 2., $R_{\neg\Diamond}$

$\downarrow$

4. $0r1$, $\langle p, 1\rangle$ — 2., $R_{\Diamond}$

$\downarrow$

5. $\langle\neg\Box p, 1\rangle$ — 3., 4., $R_{\Box}$

$\downarrow$

6. $\langle\Diamond\neg p, 1\rangle$ — 5., $R_{\neg\Box}$

$\downarrow$

7. $1r2$, $\langle\neg p, 2\rangle$ — 6., $R_{\Diamond}$

$\downarrow$

8. $0r2$ — 4., 7., Rule for transitivity

$\downarrow$

9. $\langle\neg\Box p, 2\rangle$ — 3., 8., $R_{\Box}$

$\downarrow$

10. $\langle\Diamond\neg p, 2\rangle$ — 9., $R_{\neg\Box}$

$\downarrow$

11. $2r3$, $\langle \neg p, 3 \rangle$ 10., $R_{\Diamond}$

$\downarrow$

12. $0r3$ 8., 11., Rule for transitivity

We can repeat the last four steps, introducing the next new index. The branch is not maximal according to Definition 1.21, because is infinite. However, we see that we used all rules that could be used. If some rule could be applied at some successive step, it was used at some stage of the branch. We then need a more general definition that captures finite, as well as infinite cases—variant two, the final version.

Having the definition of a core of rule 1.13, we can define a notion of strong similarity.

Definition 1.23 (*Strong similarity*) Let rule $R \in \mathbf{MRT}$ and $\langle X_1, \ldots, X_n \rangle \in R$, for some $n \in \mathbb{N}$, and $W \subseteq \mathsf{Ex}$. We say that W is strongly similar to a set X_i, where $1 < i \leq n$, iff for some $\langle Y_1, \ldots, Y_n \rangle$ that is a core of R in $\langle X_1, \ldots, X_n \rangle$:

1. W is similar to X_i
2. for some subset $W' \subseteq W$, $Y_1 \subseteq W'$
3. W' is similar to $Y_1 \cup (X_i \setminus X_1)$.

Definition 1.24 (*Maximal branch*) Let $\phi : K \longrightarrow P(\mathsf{Ex})$ be a branch. We say that ϕ is maximal iff satisfies one of the conditions:

1. ϕ is closed
2. for any rule $R \in \mathbf{MRT}$, any $n \in \mathbb{N}$, and any $X, X_1, \ldots, X_n \subseteq \mathsf{Ex}$, if:
   - $n \geq 1$
   - $\langle X, X_1, \ldots, X_n \rangle \in R$
   - $X \in \phi$

   then there is $Y_j \in \phi$, for some $j \in \mathbb{N}$, and for some $1 \leq i \leq n$, there is a set $W \subseteq \mathsf{Ex}$ strongly similar to X_i, such that $W \subseteq Y_j$.

Therefore, a maximal branch must be either closed or closed under rules (all possible rules have been applied). A maximal branch can be finite or infinite. If a branch is maximal in the sense of the former Definition 1.21, then it is also maximal under the latter definition.

By Definitions 1.19 and 1.24 we have an obvious corollary.

Corollary 1.25 *All closed branches are maximal.*

In our metatheory one of the most important notions is a notion of branch consequence relation:

Definition 1.26 (*Branch consequence relation*) Let $X \subseteq \mathsf{For}$, $A \in \mathsf{For}$. We say that A is branch consequence of X modulo **MRT** (in short: $X \rhd_{\mathbf{MRT}} A$) iff there exists some finite $Y \subseteq X$, such that all maximal branches starting from a set $\{\langle B, i \rangle : B \in Y \cup \{\neg A\}\}$, for some $i \in \mathbb{N}$, are closed.

1.5 Modal Tableaus

We intend to describe a general relationship between $\models_{\mathbf{M}}$ and $\rhd_{\mathbf{MRT}}$, for any class of models **M** and any class of rules **MRT**. In our approach tableaus are only a way of choosing a sufficient number of branches to confirm the fact that $X \rhd_{MRT} A$, for some $X \subseteq \mathsf{For}$, $A \in \mathsf{For}$. Now, we come to descriptions of tableaus.

We need an additional definition:

Definition 1.27 (*Branch maximal in a set of branches*) Let Φ be a set of branches and $\psi \in \Phi$. We say that ψ is maximal in the set Φ (in short: Φ–maximal) iff there is no branch $\phi \in \Phi$ such that $\psi \subset \phi$.

A tableau is a set of branches, so it is a more complex object than a branch. A tableau is a set of branches: (i) starting with the same set of expressions, and (ii) if there exists a branching point, then a proper **MRT**-rule was applied.

Definition 1.28 (*Tableau*) Let $X \subseteq \mathsf{For}$, $A \in \mathsf{For}$ and Φ be a set of branches. An ordered triple $\langle X, A, \Phi\rangle$ we call a tableau for pair $\langle X, A\rangle$ (or shortly: tableau) iff the following conditions are satisfied:

1. Φ is a non-empty set of branches
2. there is an index $i \in \mathbb{N}$ such that for any branch $\psi \in \Phi$, $\psi(1) = \{\langle B, i\rangle : B \in X \cup \{\neg A\}\}$
3. each branch belonging to Φ is Φ–maximal
4. for any $n, i \in \mathbb{N}$ and any branches $\psi_1, \dots, \psi_n \in \Phi$, if:
    - i and $i + 1$ belong to domains of $\psi_1, \dots, \psi_n$
    - for any $1 < k \leq n$ and $o \leq i$, $\psi_1(o) = \psi_k(o)$

   then there exists a rule $R \in \mathbf{MRT}$ and m-tuple $\langle Y_1, \dots, Y_m\rangle \in R$, where $1 < m$, that for any $1 \leq k \leq n$:

    - $\psi_k(i) = Y_1$
    - and there exists such $1 < l \leq m$ that $\psi_k(i + 1) = Y_l$.

Now, having a general notion of a tableau, we can define complete and incomplete tableaus. Intuitively, a tableau is complete iff all branches that are contained in the tableau are maximal and horizontally all possible branches are added.

Definition 1.29 (*Complete/incomplete tableau*) Let a triple $\langle X, A, \Phi\rangle$ be a tableau. We say that $\langle X, A, \Phi\rangle$ is complete iff the following conditions are satisfied:

1. every branch belonging to Φ is maximal[6]
2. there is no set of branches Ψ and tableau $\langle X, A, \Psi\rangle$, such that $\Phi \subset \Psi$.

A tableau is incomplete iff is not complete.

---

[6] This condition is not redundant. In the general definition of a tableau 1.28 every branch must be Φ-maximal (according to Definition 1.27), here must be just maximal.

Intuitively, a tableau is closed iff all branches are closed and horizontally all possible branches are added, so if it is a complete tableau with closed branches.

Definition 1.30 (*Closed/open tableau*) Let $\langle X, A, \Phi \rangle$ be a tableau. We say that $\langle X, A, \Phi \rangle$ is closed iff it satisfies the conditions:

1. $\langle X, A, \Phi \rangle$ is a complete tableau
2. every branch belonging to Φ is closed.

A tableau is open iff is not closed.

From Definition 1.30 we obtain the following immediate corollary:

Corollary 1.31 *All closed tableaus are complete.*

2 Modal Tableau Lemmas

To prove that the relations $\models_{\mathbf{M}}$ and $\rhd_{\mathbf{MRT}}$ cover the same set of pairs, we must define conditions that **M** and **MRT** should satisfy. Moreover, we need to prove some helpful facts that we will use in the proof of Tableau Metatheorem.

2.1 Lemma on Open Tableaus

Lemma 2.1 (On open tableaus) *Let **MRT** be a set of modal tableau rules, X be a finite subset of* For, *and* $A \in$ For. *If there is a maximal and open **MRT**–branch starting from* $\{\langle B, i \rangle : B \in X \cup \{\neg A\}\}$, *for some* $i \in \mathbb{N}$, *then all complete **MRT**-tableaus* $\langle X, A, \Psi \rangle$ *are open.*

Proof Let us take a set **MRT** and assume that for some finite $X \subseteq$ For, $A \in$ For, and $i \in \mathbb{N}$ there exists a maximal and open **MRT**-branch starting from $X^i = \{\langle B, i \rangle : B \in X \cup \{\neg A\}\}$. This branch will be denoted by ϕ.

Since the branch ϕ is maximal, so it is closed under the rules of **MRT** in such a sense that for any rule $R \in$ **MRT**, any $n \in \mathbb{N}$ and any $X \in \phi$, if $\langle X, X_1, \ldots, X_n \rangle \in R$, then exists some $Y \in \phi$, such that for $1 \leq i \leq n$, and some $W \subseteq Y$ that W is strongly similar to X_i , by 1.24.

Since the branch ϕ is open, so no member of ϕ is t-inconsistent, by 1.19.

We assume that there is a complete and closed **MRT**-tableau $\langle X, A, \Psi \rangle$.

Since the tableau $\langle X, A, \Psi \rangle$ is complete, so Ψ is a set of all such branches that together make $\langle X, A, \Psi \rangle$ be a complete tableau, by 1.29.

Since the tableau $\langle X, A, \Psi \rangle$ is closed, so all **MRT**-branches belonging to Ψ are closed, by 1.30. For some $k \in \mathbb{N}$ each of those branches:

- starts from $X^k = \{\langle B, k \rangle : B \in X \cup \{\neg A\}\}$, by 1.28
- ends with a t-inconsistent set, by 1.30.

We are going to show that in Ψ exists some open branch ψ that contradicts the assumption that $\langle X, A, \Psi \rangle$ is closed. We start some induction on the members of branches.

Let us consider the first member of all branches in Ψ. It is $Y_1 = X^k = \{\langle B, k \rangle : B \in X \cup \{\neg A\}\}$. Y_1 is similar to $X^i = \{\langle B, i \rangle : B \in X \cup \{\neg A\}\}$, by 1.9. Because ϕ is an open branch, so X^i, X^k and Y_1 are t-consistent, by 1.6. Since Ψ is a set of closed branches, so there must be a rule $R \in$ **MRT**, such that $\langle Y_1, Z_2, \ldots, Z_l \rangle \in R$, where $1 < l$ and for any $1 < j \leq l$ there is a branch in Ψ to which Z_j belongs, by 1.29.

Some Z_m—for $1 < m \leq l$—has to be t-consistent, because a set similar to Y_1 is equal to $X^i = X_1$, so by 1.12 (Closure under Similarity), (Closure under Expansion) and 1.24 there is $\langle X_1, X'_2 \ldots, X'_l \rangle \in R$, X'_m is similar to Z_m, X'_m is strongly similar to some $W \subseteq \mathsf{Ex}$, and Z_m is t-consistent as $W \subseteq U \in \phi$, for some $U \subseteq \mathsf{Ex}$, since ϕ is open and closed under rules. We denote the member Z_m by Y_2 and W by Y_2^*.

Therefore, for 1 there are $\phi', \phi'' \in \Psi$ such that:

- $Y_1 \in \phi'$
- Y_2 is a consequence of some $R \in$ **MRT** applied to Y_1 that produces the branch $\phi'' \in \Psi$.
- $Y_2 \in \phi''$
- $Y_1 \subset Y_2$
- for some $U \subseteq \mathsf{Ex}$, a set $Y_2^* \subseteq U \in \phi$, where Y_2^* is similar to Y_2.

Now, we assume that for some $n \in \mathbb{N}$ there are $\phi', \phi'' \in \Psi$ such that:

- $Y_n \in \phi'$
- Y_{n+1} is a consequence of some $R \in$ **MRT** applied to Y_n that produces the branch $\phi'' \in \Psi$.
- $Y_{n+1} \in \phi''$
- $Y_n \subset Y_{n+1}$
- for some $U \subseteq \mathsf{Ex}$, a set $Y_{n+1}^* \subseteq U \in \phi$, where Y_{n+1}^* is similar to Y_{n+1}.

Since the branch ϕ is open, so U, as well as Y_{n+1}^*, are t-consistent. As a consequence, Y_{n+1} is also t-consistent, by 1.6.

Since the tableau $\langle X, A, \Psi \rangle$ is complete and closed, so there must be a rule $R \in$ **MRT** and $\langle X_1, X_2, \ldots, X_k \rangle \in R$, where $k \geq 2$, $X_1 = Y_{n+1}$ and for any $1 < j \leq k$, X_j belongs to some branch in Ψ, by 1.29.

Some X_i—for $1 < i \leq k$—has to be t-consistent. If a set similar to X_i is not included in U, then because by 1.12 (Closure under Similarity), (Closure under Expansion) and 1.24 there is $\langle Z_1, \ldots, Z_k \rangle \in R$, where $Z_1 = U$, X_i is similar to some $W_i \subseteq Z_i$, Z_i is strongly similar to some Z'_i, and X_i is t-consistent as $Z'_i \subseteq U' \in \phi$, for some $U' \subseteq \mathsf{Ex}$, since ϕ is open and closed under rules. The member X_i we denote by Y_{n+2} and Z'_i by Y_{n+2}^*.

Hence, for any $n \in \mathbb{N}$, there are $\phi', \phi'' \in \Psi$ such that:

- $Y_n \in \phi'$
- Y_{n+1} is a consequence of some $R \in$ **MRT** applied to Y_n that produces the branch $\phi'' \in \Psi$.
- $Y_{n+1} \in \phi''$
- $Y_n \subset Y_{n+1}$
- for some $U \subseteq \mathsf{Ex}$, a set $Y^*_{n+1} \subseteq U \in \phi$, where Y^*_{n+1} is similar to Y_{n+1}.

The set of all sets Y_n we denote by **Y**. In **Y** there is contained at least one branch ψ such that for any $n \in \mathbb{N}$, if $Y_n \in \psi$, then there exists $Y_n \in \mathbf{Y}$.

Now, if $\psi \notin \Psi$, then $\langle Y, A, \Psi \rangle$ is not complete, since ψ starts with $Y^k = \{\langle B, k \rangle : B \in X \cup \{\neg A\}\}$ and $\langle X, A, \Psi \cup \{\psi\} \rangle$ is also a tableau. If $\psi \in \Psi$, then $\langle Y, A, \Psi \rangle$ is not closed, since, by 1.20, ψ is not closed. □

2.2 Rules Sound to Models

To formulate and prove our main result we must employ some more definitions. The first one concerns so called sound tableau rules **MRT** for some class of models **M**. The word "sound" means that applying the rules leads "from truth to truth".

Definition 2.1 (*Rules sound to models*) For any set of modal tableau rules **MRT** and any class of models **M**, we say that the set of rules **MRT** is sound to **M** iff for all sets $X_1, \ldots, X_i \subseteq \mathsf{Ex}$ (where $1 < i$), all models $\mathfrak{M} \in \mathbf{M}$ and all rules $R \in \mathbf{MRT}$, if:

- $\langle X_1, \ldots, X_i \rangle \in R$
- and $\mathfrak{M}$ satisfies X_1,

then $\mathfrak{M}$ satisfies X_j, for some $1 < j \leq i$.

2.3 Lemma on Maximal and Open Branch

Lemma 2.2 (Existence of maximal and open branch) *Let:*

- ***M** be a class of models*
- ***MRT** be any set of modal tableau rules that is sound to **M***
- $\mathfrak{M} = \langle W, R, V, w \rangle \in \mathbf{M}$
- *X be any finite set of formulas and $i \in \mathbb{N}$.*

*If $\mathfrak{M} \models X$, then there exists a maximal and open **MRT**-branch starting from $X^i = \{\langle A, i \rangle : A \in X\}$.*

Proof Let $\mathfrak{M} \models X$. So, by definition 1.10, the model $\mathfrak{M} = \langle W, R, V, w \rangle$ satisfies the set of expressions $X^i = \{\langle A, i \rangle : A \in X\}$, since for all $A \in X$, $\langle W, R, V, f(i) \rangle \models A$, where $f(i) = w$.

We denote the set of all branches starting from X^i by **X**. We know that either (1) all branches in **X** are finite, or (2) not all branches in **X** are finite.

We start from the case (1), assuming all branches in **X** are finite.

Now, we consider a branch starting from X^i—it is 1-member long and open, by 1.11. The branch is either maximal, or not. If it is maximal, so the lemma is proved. If it is not maximal, then by 1.24 there exists such a rule $R \in$ **MRT** and some n-tuple $\langle X_1, \ldots, X_n \rangle \in R$, that $X_1 = X^i$ and $\mathfrak{M}$ satisfies X_j, for some $1 < j \leq n$, because **MRT** is a set of modal tableau rules sound to the class of model **M** and $\mathfrak{M} \in$ **M**. The string X_1, X_j is an open branch, by 1.11.

We assume now, that there exists some open branch starting with X^i. It is m-member long, but not maximal. By 1.24 there is such a rule $R \in$ **MRT** and some n-tuple $\langle X_1, \ldots, X_n \rangle \in R$, that $X_1 = X_m$ and $\mathfrak{M}$ satisfies X_j, for some $1 < j \leq n$, because **MRT** is a set of modal tableau rules sound to the class of model **M** and $\mathfrak{M} \in$ **M**. Hence, the branch $X_1, \ldots, X_{m+1}$ is open, by 1.11.

(†) In consequence, for any $n \in \mathbb{N}$ and any branch ϕ such that $\phi(1) = X_1 = X^i$, if ϕ is n-member long and not maximal, then for some branch ψ, $\phi \subset \psi$, ψ is $n + 1$-member long and ψ is open.

We assume towards contradiction that no branch of **X** is maximal. However, by (†) it means that there exists an infinite branch: $X_1, X_2, \ldots,$ where

1. $X_1 = X^i$
2. X_{n+1} is the last set of an open branch $\phi \in$ **X** such that:
   a. ϕ is n-member long
   b. for all $i \leq n$, some $X_n \in \phi$.

But this is in contradiction with the fact that all branches in **X** are finite.

Now, we begin the case (2), assuming that not all branches in **X** are finite. Let all finite branches in **X** be closed.

Let $Y \subseteq \mathsf{Ex}$ be a finite set of expressions. By 1.15, the number of rules that can be applied to Y is finite. So, the set $\mathbf{MRT}_Y$ has j members, for some $j \in \mathbb{N}$. Each of the rule in $\mathbf{MRT}_Y$ we denote by R^i $(1 \leq i \leq j)$. By 1.15, for any $R^i \in \mathbf{MRT}_Y$, there is a finite number of n-tuples $\langle Y, X_1, \ldots, X_{n-1} \rangle$ in R^i. Hence, any R^i_Y, where $R^i \in \mathbf{MRT}_Y$, is finite. Each of n-tuple in any R^i_Y we denote by r_k $(1 \leq k)$. So in any R^i_Y, where $R^i \in \mathbf{MRT}_Y$, there is a finite number of n-tuples $\langle Y, X_1, \ldots, X_{n-1} \rangle$, that can be listed: $r^i_1, \ldots, r^i_k$, where $1 \leq k$.

Now, we make a list of all r^i_l in all R^i_Y, giving them some type of lexicographical order: $\underbrace{r^1_1, \ldots, r^1_m}_{R^1_Y}, \underbrace{r^2_1, \ldots, r^2_n}_{R^2_Y}, \ldots, \underbrace{r^j_1, \ldots, r^j_o}_{R^j_Y}$, where $1 \leq m, n, o$.

Any list of n-tuples for Y we call a *Y-list* and denote by L_Y. Of course, there can be many lists for Y.

Having some L_Y and some $r_i \in L_Y$, we know that $r_i \in R^k_Y \subseteq R^k$, for some $k \leq j$. Saying that r_j is an *expansion* of r_i we mean that:

- $r_j \in R^k$
- $r_i = \langle X_1, \ldots, X_n \rangle$, $r_j = \langle Z_1, \ldots, Z_n \rangle$ and for any $1 \leq l \leq n$, $Z_1, \ldots, Z_n$ are sets that satisfy the conditions:

  1. $X_l \subset Z_l$.
  2. X_i is similar to $X_1 \cup (Z_i \setminus Z_1)$.

If r_j is a given *expansion* of r_i, we will write r'_i instead of r_j.

(∗) By the definition of tableau rule 1.12 (Closure under Expansion), we know that for any $r_i = \langle X_1, \ldots, X_n \rangle$ belonging to some rule R and any Z_1, if Z_1 is t-consistent, $X_1 \subset Z_1$, and for all $1 < i \leq n$, X_i is not similar to any subset of Z_1, then there is exactly one r_j that is an expansion of r_i, where $r_j = \langle Z_1, \ldots, Z_n \rangle$, for some $Z_2, \ldots Z_n \subseteq \mathsf{Ex}$.

Let L_Y be some Y-list. By induction we define a closure of Y under L_Y. $L_Y(Y)$ is a maximally long string $Z_1, \ldots, Z_o$, for some $o \in \mathbb{N}$, where for any $1 \leq n \leq o$:

1. if $n = 1$, then $Z_n = Y$
2. if $n = 2$, then $Z_n = X_j$, where

   a. r_1 is the first n-tuple in L_Y
   b. $r_1 = \langle Y, X_1, \ldots, X_j, \ldots, X_n \rangle$, for $1 \leq n$
   c. X_j is some t-consistent set in r_1 different from Y

3. if $n > 2$ and

   a. Z_{n-1} exists in the string
   b. Z_{n-1} is t-consistent
   c. Z_{n-1} is a consequence of an expansion of some m-tuple $r_l \in L_Y$ applied to Z_{n-2}, so $r'_l = \langle Z_{n-2}, W_1, \ldots, W_m \rangle$, for some $1 < m$ and $Z_{n-1} = W_k$, for $1 < k \leq m$

   then $Z_n = X_j$, where:

   a. there is r_{l+n} and it is the first k-tuple in L_Y, where $1 \leq n$ such that
   b. r'_{l+n} is an expansion of r_{l+n}
   c. $r'_{l+n} = \langle Z_{n-1}, X_1, \ldots, X_j, \ldots, X_i \rangle$, for $1 \leq i$ and $1 \leq j \leq i$
   d. X_j is some t-consistent set in r'_{l+n} different from Z_{n-1}.

By definition of branch 1.16 any $L_Y(Y) := Z_1, \ldots, Z_n$, for some $n \in \mathbb{N}$, is a branch. Moreover, by the assumption that the rules of **MRT** are sound to **M**, if $\mathfrak{M} \in \mathbf{M}$ and $\mathfrak{M}$ satisfies Z_1, then for some $L_Y(Y) := Z_1, \ldots, Z_n$, $\mathfrak{M}$ satisfies Z_n.

Now, we take our initial set of expressions X^i and conclude:

- X^i is finite, so we have some branch $L^1_{X^i}(X^i) := X_1, \ldots, X_k$, for some X^i-list and some $k \in \mathbb{N}$, such that:
- since for some model $\mathfrak{M} \in \mathbf{M}$, $\mathfrak{M}$ satisfies X^i, so $\mathfrak{M}$ satisfies also X_k and X_k is t-consistent

- by the assumption (2), $L^1_{X^i}(X^i)$ is not a maximal branch, since there is no finite, open and maximal branch in **X**
- since rules of **MRT** are closed under finite sets 1.12, hence X_k is finite, too.

We consider the string of closures under some lists L^j — where $j \in \mathbb{N}$—and assume that set X_o is finite and $\mathfrak{M}$ satisfies X_o:

$$L^1_{X^i}(X^i) := X_1, \ldots, X_k \qquad \text{for some } k > 1 \in \mathbb{N}$$

$$L^2_{X_k}(X_k) := X_k, \ldots, X_l \qquad \text{for some } l > k \in \mathbb{N}$$

.

.

.

$$L^{j-1}_{X_{l+m}}(X_{l+m}) := X_{l+m}, \ldots, X_n \qquad \text{for some } n \text{ and } m, n > l+m \in \mathbb{N}$$

$$L^j_{X_n}(X_n) := X_n, \ldots, X_o \qquad \text{for some } o > n \in \mathbb{N}$$

Now, we have:

- X_o is finite, so we have some branch $L^{j+1}_{X_o}(X_o) := X_o, \ldots, X_r$, for some X_o-list and some $r \in \mathbb{N}$, such that:
- Since for some model $\mathfrak{M} \in \mathbf{M}$, $\mathfrak{M}$ satisfies X_o, so $\mathfrak{M}$ satisfies also X_r and X_r is t-consistent.
- By the assumption (2) and definition 1.18, the branch $(((\ldots(L^1_{X^i}(X^i) \oplus L^2_{X_k}(X_k)) \oplus \ldots) \oplus L^{j-1}_{X_{l+m}}(X_{l+m})) \oplus L^j_{X_n}(X_n)) \oplus L^{j+1}_{X_o}(X_o)$ is not a maximal branch, since there is no finite, open and maximal branch in **X**.
- Since rules of **MRT** are closed under finite sets 1.12, hence X_r is finite, too.

As a consequence, for any $j \in \mathbb{N}$ there is an open branch $L^{j+1}_{X_m}(X_m)$, where $L^j_{X_l}(X_l) = X_l, \ldots, X_m$, $L^1_{X^i}(X^i) = X_1, \ldots, X_k$, for some $k < l < m \in \mathbb{N}$, $X_1 \subset X_k \subset \cdots \subset X_l \subset X_m$ and $\mathfrak{M}$ satisfies X_m. We can list those branches:

$$\underbrace{X_1 = X^i, \ldots, X_k}_{L^1_{X^i}(X^i)}, \underbrace{X_k, \ldots, X_l}_{L^2_{X_k}(X^k)}, \underbrace{X_l, \ldots, X_m}_{L^3_{X_l}(X^l)}, \ldots$$

and as a result, omitting double occurrences of members, we obtain an infinite branch starting from X^i: $X_1, X_2, X_3, \ldots$, by Definition 1.16. This branch is not closed, by 1.19.

Now, we check whether the branch is maximal. By 1.24 we take some rule $R \in$ **MRT** and for $1 < n \in \mathbb{N}$ some sets $Y_1, \ldots Y_n \subseteq \mathsf{Ex}$ such that:

- $\langle Y_1, \ldots, Y_n \rangle \in R$
- for some $1 \leq i$, $X_i = Y_1$

The question is whether there is $j \in \mathbb{N}$ such that for some $1 < k \leq n$, some subset of X_j is strongly similar to Y_k and X_j is in the branch.

We know that $X_i \in L^k_{X_m}(X_m)$, for some $1 \leq k$ and $m \leq i$. By 1.15, (a) $R \in \mathbf{MRT}_{X_m}$, or (b) not. If not, then there exists some X_o, where $m < o \leq l$, $L^{k+1}_{X_l}$, for some $l \in \mathbb{N}$, and $R \in \mathbf{MRT}_{X_l}$, by the construction of the X^i, its supersets-lists, and the considered branch.

Firstly, we assume that $R \in \mathbf{MRT}_{X_m}$. By construction of the X^i, its supersets-lists and the considered branch there are three possibilities:

1. $X_{i+1} = Y_k$, for some $1 < k \leq n$
2. there is an n-tuple $\langle W_1, \ldots, W_n \rangle \in R$, an expansion of $\langle Y_1, \ldots, Y_n \rangle$, $X_{i+o} = W_1$, where $o \geq 1$, $X_{i+o+1} = W_k$, and some subset of W_k is strongly similar to Y_k, for some $1 < k \leq n$
3. there is a set X_{i+o}, where $1 \leq o$, and other rule $R' \in \mathbf{MRT}_{X_m}$, such that $\langle X_{i+o-1}, Y_1, \ldots, Y_r \rangle \in R'$, for some $r \in \mathbb{N}$, $X_{i+o} = Y_{r_1}$, where $1 \leq r_1 \leq r$, and some subset of X_{i+o} is strongly similar to Y_k, for some $1 < k \leq n$.

In the case (b) we have analogical possibilities. Therefore, we have an open and maximal branch. □

2.4 *Model Generated by Branch*

Definition 2.2 (*Model generated by branch*) Let **MRT** be any set of modal tableau rules. Let ϕ be a **MRT**-branch and $X = \{\langle A, k \rangle : A \in Y\} \subseteq \bigcup \phi$, for some $k \in \mathbb{N}$ and some nonempty $Y \subseteq \mathsf{For}$. We define a set $AT(\phi)$ as follows: $x \in AT(\phi)$ iff one of the conditions holds

- $x \in \bigcup \phi \cap (\{irj : i, j \in \mathbb{N}\} \cup \{i = j : i, j \in \mathbb{N}\})$
- $x \in \bigcup \phi \cap (\mathsf{Var} \times \mathbb{N})$.

We say that $\mathfrak{M} = \langle W, R, V, w \rangle$ is a *model generated by branch* ϕ iff

- $W = \{i : i \in *(AT(\phi))\}$
- for any $i, j \in \mathbb{N}$,
  - $\langle i, j \rangle \in R$ iff $irj \in AT(\phi)$
  - if $i = j \in AT(\phi)$, then in the model i is identical to j
  - $V(x, i) = 1$ iff $\langle x, i \rangle \in AT(\phi)$
- $w = k$.

2.5 *Open and Maximal Branch Generates Model*

Corollary 2.3 (Open and maximal branch generates model) *Let ϕ be an open and maximal **MRT**-branch, for some set of modal tableau rules **MRT**. Let $X = \{\langle A, k \rangle : A \in Y\} \subseteq \bigcup \phi$, for some $k \in \mathbb{N}$ and some nonempty $Y \subseteq \mathsf{For}$. Then there exists a model $\mathfrak{M}$ generated by ϕ.*

Proof By definitions of open branch 1.19, maximal branch 1.24 and model generated by branch 2.2. □

2.6 Models Sound to Rules

To formulate and prove Tableau Metatheorem we must employ one more definition. It concerns so-called sound class of models **M** to some set of modal tableau rules **MRT**. In practice, it means that if applying the rules generates a model, then it belongs to **M** and the model satisfies an initial set of expressions.

Definition 2.4 (*Models sound to rules*) Let

- **MRT** be any set of modal tableau rules
- ϕ be any maximal **MRT**-branch and $X \times \{i\} \subseteq \cup\phi$, for some $X \subseteq \mathsf{For}$ and some $i \in \mathbb{N}$
- **M** be any class of models.

We say that **M** is sound to rules of **MRT** iff for all models $\mathfrak{M}$ generated by ϕ:

- $\mathfrak{M} \in \mathbf{M}$
- $\mathfrak{M} \models X$.

2.7 Closure Under Rules

Definition 2.5 (*Closure under rules*) Let $X \subseteq \mathsf{Ex}$. We say that $Y \subseteq \mathsf{Ex}$ is a closure of X under **MRT** iff Y is a set that satisfies the conditions

- $X \subseteq Y$
- for any rule R of **MRT** and any n-tuple $\langle Z_1, Z_2, \ldots, Z_n\rangle \in R$, where $n \in \mathbb{N}$, if $X \subseteq Z_1 \subseteq Y$, then $Z_j \subseteq Y$, for some $2 \leq j \leq n$.

Any set Y that is a closure of X under **MRT** we denote by $\mathbf{MRT}_{(X)}(Y)$.

For any set of expressions we have at least one closure under rules, but sometimes there can be more closures.

Lemma 2.3 (On existence of open and maximal branch) *Let $X \subseteq \mathsf{For}$ and $i \in \mathbb{N}$. If for all finite $Y \subseteq X$ exists a maximal and open branch starting with $Y^i = \{\langle A, i\rangle : A \in Y\}$, then there is some closure of $X^i = \{\langle A, i\rangle : A \in X\}$ under **MRT** that is an open and maximal branch.*

Proof We take any $X \subseteq \mathsf{For}$, $i \in \mathbb{N}$, and assume that $(*)$ for any finite subset $Y \subseteq X$ there exists an open and maximal branch starting from a set of expressions $Y^i = \{\langle A, i\rangle : A \in Y\}$.

We take the set of all maximal and open branches that start from a set $Y^i = \{\langle A, i\rangle : A \in Y\}$, for any finite $Y \subseteq X$. We denote the set by $\mathbf{X}$.

Now, we define a set $\overline{\mathbf{X}}$ with the conditions:

1. $\overline{\mathbf{X}} \subseteq \mathbf{X}$
2. for every two branches ϕ and ψ in $\mathbf{X}$, if there exist such two numbers $i, k \in \mathbb{N}$ that $\phi(i) \cup \psi(k)$ is a t-inconsistent set, then $\phi \notin \overline{\mathbf{X}}$ or $\psi \notin \overline{\mathbf{X}}$
3. $\overline{\mathbf{X}}$ is a maximal set among those subsets of $\mathbf{X}$ that satisfy conditions 1. and 2.

There exists at least one set $\overline{\mathbf{X}}$ such that $\emptyset \subset \overline{\mathbf{X}} \subseteq \mathbf{X}$. We take one such set, denoting it as $\overline{X}$.

We consider the set $\bigcup\{\phi(1) : \phi \in \overline{X}\}$. We observe that (∗∗) $X^i \subseteq \bigcup\{\phi(1) : \phi \in \overline{X}\}$. If $X^i \nsubseteq \bigcup\{\phi(1) : \phi \in \overline{X}\}$, then there exists $x \in X^i$, $x \notin \bigcup\{\phi(1) : \phi \in \overline{X}\}$ and for any branch $\psi \in \mathbf{X}$, if $x \in \psi(1)$, $\psi(1) \subseteq X^i$ and $\psi(1)$ is finite, then $\psi \notin \overline{X}$. But then for all finite sets $Y^i \subseteq X^i$ there is no maximal and open branch starting from a set $Y^i \cup \{x\}$, which contradicts the assumption (∗).

We put a condition:

$U \in \overline{\overline{X}}$ iff there exists such a branch ϕ that $\phi \in \overline{X}$ and $U = \bigcup \phi$

defining a set $\overline{\overline{X}}$. Now, we can define a set $Z = \bigcup \overline{\overline{X}}$.

We claim that Z is a closure of $X^i = \{\langle A, i\rangle : A \in X\}$ under **MRT** (Definition 2.5) and Z is an open and maximal branch.

Firstly, we show that Z is a closure of $X^i = \{\langle A, i\rangle : A \in X\}$, so it satisfies the conditions of 2.5.

Let us observe that $X^i \subseteq Z$, since (∗∗) $X^i \subseteq \bigcup\{\phi(1) : \phi \in \overline{X}\}$, and by construction of Z, $\bigcup\{\phi(1) : \phi \in \overline{X}\} \subseteq Z$.

We take any rule $R \in$ **MRT** and n-tuple $\langle U_1, \ldots, U_n\rangle \in R$, for some $n \in \mathbb{N}$, and assume that $X^i \subseteq U_1 \subseteq Z$. By Definition 1.12 it follows that there exists such n-tuple $\langle U_1', \ldots, U_n'\rangle \in R$ that:

- for any $1 \leq j \leq n$, U_j' is a minimal, finite set that, if U_j is not a minimal, finite set such that $\langle U_1, \ldots, U_n\rangle \in R$, then $U_j' \subset U_j$
- for any $1 < j \leq n$, $U_j \setminus U_1 = U_j' \setminus U_1'$.

In consequence, assuming that $U_1' \subseteq Z$, we must show that for some $1 < l \leq n$, $U_l' \subseteq Z$, since $U_l' \cup U_1 = U_l$. Because for the finite set of expressions U_1' it is the case that $U_1' \subseteq Z$, so there exists a finite number of branches $\phi_1, \phi_2, \ldots, \phi_o$ in $\overline{X}$ and for some $k \in \mathbb{N}$, $U_1' \subseteq \phi_1(k) \cup \phi_2(k) \cup \cdots \cup \phi_o(k)$. Hence, to the set $\overline{X}$ belongs such a branch ψ that $\psi(1) = \phi_1(1) \cup \phi_2(1) \cup \cdots \cup \phi_o(1)$ i $U_1' \subseteq \psi(m)$, for some $m \in \mathbb{N}$, and since ψ is a maximal branch, $\phi_1(k) \cup \phi_2(k) \cup \cdots \cup \phi_o(k)$ is t-consistent, so—by Definition 1.24—for some $1 < l \leq n$, $U_l' \subseteq \bigcup \psi$. In consequence $U_l' \subseteq Z$, because by construction of Z, $\bigcup \psi \subseteq Z$.

Now, we show that Z is an open and maximal branch. We know—by definition of branch 1.16—that Z is a branch.

By construction of Z, Z is an open branch, i.e. no subset of Z is t-inconsistent, by definition of $\overline{X}$.

We check now, if Z is a maximal branch. In the light of the definition of maximal branch 1.24 we assume that there is a rule $R \in$ **MRT** and n-tuple $\langle X_1, \ldots, X_n\rangle \in R$, for some $n \in \mathbb{N}$, where $X_1 = Z$. By definition of tableau rules 1.12 there exists such n-tuple $\langle X'_1 \ldots, X'_n\rangle \in R$ that for any $1 < j \leq n$, $X_j \setminus X_1 = X'_j \setminus X'_1$ and $X^i \subseteq X'_1 \subseteq Z$. Since Z is a closure of X^i, so $X'_j \subseteq Z$, for some $1 < j \leq n$, by Definition 2.5. Hence $X_j \subseteq Z$, because $X_j = X_1 \cup X'_j$. But then $X_1 \not\subset X_j$, which by definition of tableau rules 1.12 is impossible. In consequence there is not a tableau rule and n-tuple $\langle X_1, \ldots, X_n\rangle \in R$, where $X_1 = Z$, for some $n \in \mathbb{N}$. Therefore Z is a maximal branch, by definition of maximal branch 1.24. □

2.8 Modal Tableau Metatheorem

From those definitions and lemmas we can conclude Tableau Metatheorem for modal logics defined by possible worlds' semantics:

Theorem 2.6 (Modal Tableau Metatheorem) *For any set of modal tableau rules* ***MRT*** *and any class of models* $\mathbf{M}$, *if:*

1. *set of rules* ***MRT*** *is sound to class of models* $\mathbf{M}$
2. *class of models* $\mathbf{M}$ *is sound to rules of* ***MRT***

then for all $X \subseteq For$, $A \in For$ *the following statements are equivalent:*

- $X \models_M A$
- $X \rhd_{MRT} A$
- *there is a finite* $Y \subseteq X$ *and a closed tableau* $\langle Y, A, \Phi\rangle$.

Proof We assume the points 1., 2. and take any $X \subseteq For$, $A \in For$. We must show that three implications hold.

(a) $X \models_\mathbf{M} A \Longrightarrow X \rhd_{MRT} A$

We assume $X \not\rhd_{MRT} A$. Hence, for any finite $Y \subseteq X$ there exists some branch that starts with $\{\langle B, i\rangle : Y \cup \{\neg A\}\}$—for some $i \in \mathbb{N}$—which is maximal and open, by 1.26. By Lemma 2.3 (On existence of open and maximal branch), there exists some closure of $\{\langle B, i\rangle : B \in X \cup \{\neg A\}\}$ under **MRT** that is a maximal and open branch ψ. By the Collorary 2.3, Definition 2.4 and point 1., we know that there is a model $\mathfrak{M} \in \mathbf{M}$ generated by ψ and $\mathfrak{M} \models X \cup \neg A$. So, $\mathfrak{M} \models X$ and $X \not\models A$, by 1.1. As a consequence $X \not\models_\mathbf{M} A$.

(b) $X \rhd_{MRT} A \Longrightarrow$ there is a finite $Y \subseteq X$ and a closed tableau $\langle Y, A, \Phi\rangle$

We assume that for any finite $Y \subseteq X$ all tableaus $\langle Y, A, \Phi\rangle$ are open. We take some complete tableau $\langle Y_0, A, \Phi\rangle$, for a finite $Y_0 \subseteq X$. Under the last assumption the tableau $\langle Y_0, A, \Phi\rangle$ is open, too.

If $\langle Y_0, A, \Phi\rangle$ is open and complete, then to Φ belongs a maximal and open branch ϕ that starts with $\{\langle B, i\rangle : B \in Y_0 \cup \{\neg A\}\}$, for any $i \in \mathbb{N}$.

Since Y_0 is any finite subset of X, so for any finite $Y \subseteq X$ there is some branch ψ starting with some $\{\langle B, i\rangle : B \in Y \cup \{\neg A\}\}$, for any $i \in \mathbb{N}$ that is maximal and open.

In consequence, there is no such finite subset $Y \subseteq X$ that all maximal branches starting with $\{\langle B, i\rangle : B \in Y \cup \{\neg A\}\}$, for any $j \in \mathbb{N}$, are closed. Therefore, $X \not\triangleright_{MRT} A$.

(c) There is a finite $Y \subseteq X$ and a closed tableau $\langle Y, A, \Phi\rangle \Longrightarrow X \models_{\mathbf{M}} A$.

We assume that $X \not\models_{\mathbf{M}} A$. Hence, there is such a model $\mathfrak{M} \in \mathbf{M}$ that $\mathfrak{M} \models X$ and $\mathfrak{M} \not\models A$. Consequently, $\mathfrak{M} \models X \cup \{\neg A\}$, so for any finite $Y \subseteq X, \mathfrak{M} \models Y \cup \{\neg A\}$.

We take a finite $Y_0 \in X$. From the lemma 2.2 (Existence of maximal and open branch) and point 2. we obtain a corollary that for any $i \in \mathbb{N}$ there exists a maximal and open branch starting with $\{\langle B, i\rangle : B \in Y_0 \cup \{\neg A\}\}$.

Therefore by 2.1 (On open tableaus) each tableau $\langle Y_0, A, \phi\rangle$ is open. Since Y_0 is any finite subset of X, so there is no finite $Y \subseteq X$ and closed tableau $\langle Y, A, \Phi\rangle$. □

3 Summary

According to metatheorem 2.6, if for a class of models **M** a set of rules **MRT** satisfying conditions 1 and 2 is defined, then in consequence we obtain the complete and sound modal tableau system.

The formal theory presented in the paper offers a simplification of a process of defining all notions and proving particular facts while constructing a modal tableau system. What is covered by the theory turns out to be all general features of any modal tableau system determined by possible world semantics. Moreover it allows to define suitably some set of modal tableau rules in such a way that the sufficient condition for completeness and soundness of the system is satisfaction of the aforementioned conditions. In the standard approach—in contrast to the one presented—it seems to be very difficult to prove general facts about the classes of logics, since we do not have universal and precise notions that are constant and vary only from one set of tableau rules to another.

A natural problem that emerges in this context is whether our approach is applicable to strong/weak completeness with respect to structures. This problem seems to be even more interesting in light of the fact that some traditional axiomatic structures are only weekly complete with respect to certain classes of structures.[7]

In our opinion such generalizations are possible, but require further research and we treat this work as a starting point towards them. Obviously, this kind of approach can be extended to other types of logics, in order to capture relations between tableau rules and models, since the metatheory contains no limiting features that could narrow the scope of application to the range of modal logic only.

[7] See Chap. 4.4 of [3].

References

1. Jarmużek, T. (2008). Tableau system for logic of categorial propositions and decidability. *Bulletin of the Section of Logic*, *37*(3/4), 223–231.
2. Jarmużek, T. (2013). *Formalizacja metod tablicowych dla logik zdań i logik nazw (Formalization of tableau methods for propositional logics and for logics of names)*. Toruń: Wydawnictwo UMK.
3. Blackburn, P., de Rijke, M., & Vennema, Y. (2002). *Modal logic*. Cambridge.
4. Rajeev, G. *Tableau methods for modal and temporal logics* (pp. 297—396) (in [1]).

On the Essential Flatness of Possible Worlds

Neil Kennedy

Abstract The objective of this paper is to introduce and motivate a new semantic framework for modalities. The first part of the paper will be devoted to defending the claim that conventional possible worlds are ill-suited for the semantics of certain types of modal statements. We will see that the source of this expressive limitation comes from what will be dubbed "worldly flatness", the fact that possible worlds don't determine modal facts. It will be argued that some modalities are best understood as quantifiers over modal facts and that possible worlds semantics cannot achieve this. In the second part of the paper, I will present a new semantic framework that allows for such an understanding of modalities.

Keywords Possible worlds semantics · Epistemic logic · Multidimensional frames

1 Flat Worlds

The basic observation underlying the work in this paper is that possible worlds are flat: no possible world, as given by Kripke semantics, determines a modal fact. I take this to be very uncontroversial claim, but since it is a crucial one to the present analysis I will spell it out in more detail.

Recall that the celebrated notion of a Kripke model is a tuple $\langle W, R, I\rangle$ consisting of three things: (i) a set W of possible worlds, (ii) an accessibility relation R on W, and (iii) an interpretation function I assigning a set $I(p)$ of worlds to each atomic statement p. The set of worlds W is basically a basin of variation, where each element in W can be understood as a way things could be in the "universe" under consideration. R is an accessibility relation on W, one that essentially determines all the modal properties of the model. And, for each atomic statement p, the set $I(p)$ is the proposition p defines. The binary relation R is the repository of all the

N. Kennedy (✉)
Université du Québec à Montréal, Montreal, Canada
e-mail: neil.patrick.kennedy@gmail.com

R. Ciuni et al. (eds.), *Recent Trends in Philosophical Logic*,
Trends in Logic 41, DOI: 10.1007/978-3-319-06080-4_9,

fundamental modal facts of M. Without R, W and I can only determine the truth conditions of the atomic statements at each world (and their Boolean combinations); with R, the truth values of *all* the statements can be given. Most importantly, and central to the present paper, no possible world determines the binary relation R in any way, no world determines the fundamental modal features in the model. This is what worldly flatness consists in.[1]

In a nutshell, I will argue that flat worlds translate into expressive limitations for Kripke semantics. The truth conditions of certain types of modal discourse would be considerably simplified if the modal facts were directly located in the world themselves, rather than the overlying structure. A proper semantic account for this modal discourse would have worlds (or parts of these worlds) that determine accessibility relations of a certain kind.

2 Worlds with Modal Parts

In many ways, the semantic framework I will be proposing can be seen as a generalization of Prior's "Ockhamist" semantics ([9], Chap. 7). It will therefore be enlightening to examine the latter in order to arrive at a general idea of what will be put forward later on.

Prior develops a semantic framework for tense and possibility in the context of a discussion of the problem of future contingents. The problem of future contingents, as he sees it, can be understood as the problem of reconciling two seemingly incompatible properties of time. On the one hand, tense operators presuppose a linear conception of time, but on the other, possibility presupposes forward temporal branching.

Prior's solution to this problem is to think of the truth of a statement as relative not only to a moment in time but also to a world history. Relative to both a time and history, a tensed statement has a determinate truth value; relative only to a time, it can sometimes lack one. The entities at which statements are evaluated are thus moment-history pairs, a moment being a sort of snapshot of the universe and a history a sequence of such snapshots.

Formally, Prior employs the notion of an *Ockhamist model* to express this idea ([9], p. 126). An Ockhamist model is basically a tree $\mathscr{T} = \langle T, < \rangle$.[2] T is the set of all moments and $<$ the relation of temporal antecedence. Histories are construed as paths in $\mathscr{T}$ of maximal length. If, at a given moment $t \in T$, two histories h and g share the same past up to (and including) t, then both histories are possible from one another at t. The meanings of the tense and metaphysical modalities in an Ockhamist

[1] The term "flat" is sometimes used to mean that the modal facts supervene on the basic facts, as Humean supervenience would have it (cf. [8], p. 14). It seems that in this usage of the term, "flat" applies to the universe, whereas my flatness applies to single worlds.

[2] $\mathscr{T}$ is a tree iff $<$ is an anti-reflexive, transitive and connected relation $<$ on T such that, for all $t \in T$, its restriction to $\{s \in T : s \leq t\}$ is linear. In a tree, there is only one path from right to left but possibly many from left to right.

model are then defined in the following way. Relative to a history h and at moment t: (a) $P\phi$ is true iff there is a moment s of h occurring before t where ϕ is true, (b) $F\phi$ is true iff there is a moment s of h occurring after t where ϕ is true, and (c) $\Diamond\phi$ is true iff there is history g possible from h at t where ϕ is true relative to g and at t.

The point worth highlighting here is that Priorean semantics uses a notion of possible world that has modal parts. By possible world, I mean the thing at which we evaluate statements, which in this case is a pair (t, h). The observation is that h is a modal part of the possibility (t, h); h describes the way in which t stands in respect to other moments, and it does so by determining an accessibility relation on the set of moments T (the binary relation in question being $< \cap (h \times h)$).

Our own semantic proposal will retain this overall aspect of Prior's idea. In this framework, a possibility will be understood as a tuple of coordinates, each coordinate corresponding to the value of a basic state type of the universe. Just like histories, some of these states will be *modal* states in the sense that they will determine modal facts about the universe; more precisely, this means that each modal state will determine an accessibility relation on a certain sub-product of the set of possibilities.

Let us illustrate how this general idea works out in a particular case. Suppose that a universe U consists of an epistemic agent and a collection of objects that admit properties like colour, shape, weight, size, etc. The idea is that U could be nicely "factored" into three distinct components. First, a "worldly" state type $\mathtt{f}$ would determine the worldly facts, with each worldly state $f \in W_{\mathtt{f}}$ corresponding to a distribution of first-order properties on the objects of U. Second, an epistemic state type $\mathtt{e}$ would determine the epistemic facts, with each epistemic state $e \in W_{\mathtt{e}}$ corresponding to the knowledge the agent has of the objects in the world. Technically speaking, each $e \in W_{\mathtt{e}}$ would determine an epistemic accessibility relation on $W_{\mathtt{f}}$. Finally, an alethic state type $\mathtt{p}$ would determine the type of possibility in U, with each modal state $p \in W_{\mathtt{p}}$ corresponding to a kind of possibility, e.g., physical, metaphysical, conceivable, etc. Technically speaking, each element of $W_{\mathtt{p}}$ would determine an accessibility relation on $W_{\mathtt{f}} \times W_{\mathtt{e}}$. According to this picture, a possibility $\mathbf{w}$ is given as a triplet (f, e, p), with $f \in W_{\mathtt{f}}$, $e \in W_{\mathtt{e}}$ and $p \in W_{\mathtt{p}}$, it is a specific "position" in the space generated by these state types.

3 The Semantic Inconvenience of Flatness

What is the need for a notion of possibility that has modal parts? The idea is to have a simple means for representing variation in modal facts. We will examine in this section a few examples that I claim will show the necessity for such possibilities. This by no means entails that Kripkean possible worlds semantics *cannot* be used to represent such variation; however, we will see that it does so very poorly.

Variation in modal facts is especially useful for expressing the meanings of modalities M_1 and M_2 that, like tense and possibility, have some element of subordination between them. For example, if we think of knowledge and tense in the precise context where knowledge evolves in time and concerns only atemporal facts, it would seem

Fig. 1 Knowledge and tense

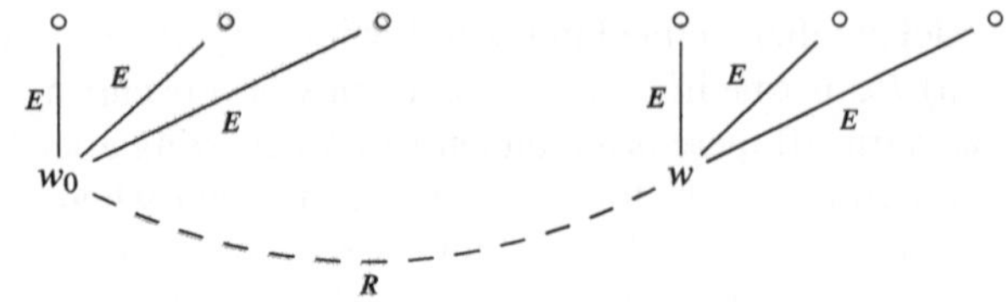

that knowledge is subordinate to tense in the following way: the world at time t is constituted in part by knowledge at time t, so knowledge is indexed by t and is part of what is being quantified over by a tense modality. When evaluating a statement of the form FKp (e.g. "At some time in the future, Mary will know that smoking is bad"), the modality F has this higher-order relation to K that makes it special. The claim here is that the ideal way of understanding F is as a quantifier over possible accessibility epistemic relations for K, each one representing Mary's knowledge state at a given time. But this is not the way it is understood in Kripke semantics.

In a Kripke model, the meanings of modalities F and K are given in terms of a frame $\langle W, R, E \rangle$, where W is a set of possible worlds (with enough variation for both modal notions), and where R and E are temporal and epistemic accessibility relations respectively. The truth of FKp at some world w_0 is understood as the existence of $w \in W$ such that w is later than w_0, i.e. $R(w_0, w)$, and Kp is true at w. In turn, Kp is true at w iff p is true at all v's accessible from w via E. The truth conditions of FKp involve no quantification over various possible accessibility relations for K. Variation in epistemic facts is accomplished by "relocating" to another region of W where E's behaviour is different, i.e. where Mary knows that smoking is bad. We can picture the idea with the following diagram: In the region surrounding w_0, E can "see" worlds in which smoking is good; but in the region surrounding w, E only sees worlds where smoking is bad. E is not just one epistemic profile but many; in fact, it comprises all of Mary's epistemic profiles across all times (Fig. 1).

Such examples are not hard to come by as they are fairly commonplace. In the remainder of this section, we will show that epistemic modalities exhibit a similar behaviour, and will thereby also benefit from a semantics with explicit modal facts. The reason for this is that epistemic modalities pertain to a very broad array of facts, since we can have knowledge (and ignorance) about virtually anything. Some of these facts will themselves be epistemic in nature, as in the case of knowledge of knowledge, be it our own knowledge or that of others. Representing knowledge of epistemic facts will require the same kind of modal variation in possible worlds. However, modal variation will be difficult to provide for using flat worlds. As a general rule, the more sophisticated the epistemic facts that are known, the harder it will be to represent this knowledge in a clear and natural way.

Consider a universe consisting solely of a circle and a square. In this universe, the circle can be either red or green, and the square either blue or yellow. Those are the only properties these individuals can have.[3] There are thus four distinct distributions

[3] Note that "being a square" and "being a circle" are not properties of this universe.

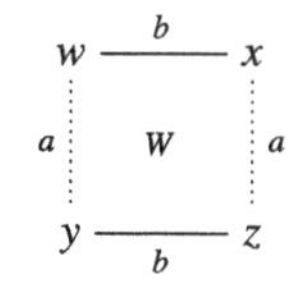

Fig. 2 Knowledge without higher-order ignorance

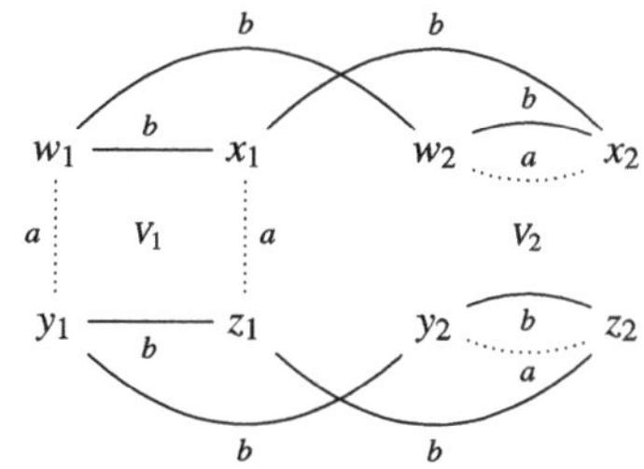

Fig. 3 Knowledge with some higher-order ignorance

of properties in this universe, one for each possible colouring of the circle and square, giving rise to four distinct possible worlds: $w = (\mathtt{r}, \mathtt{b})$, $x = (\mathtt{r}, \mathtt{y})$, $y = (\mathtt{g}, \mathtt{b})$ and $z = (\mathtt{g}, \mathtt{y})$, where the first coordinate specifies the colour of the circle and the second the colour of the square. Let us add to this universe the epistemic agents Alice and Bob, say a and b.[4] Alice is red-green colour-blind (RGC) and Bob is blue-yellow colour-blind (BYC). These epistemic states obviously give rise to the following epistemic accessibility relations on $W = \{w, x, y, z\}$: $E_a(u, v)$ iff the circle has the same colour in u and v, and $E_b(u, v)$ iff the square has the same colour in u and v. In other words, E_a connects blue and yellow worlds and E_b connects red and green ones. The resulting universe, say U, is depicted in Fig. 2.

A simple verification will show that, in the model of Fig. 2, Alice and Bob are knowledgeable about each other's colour-blindness. This simply results from the fact that there is no possibility at which Alice or Bob's colourblindness is different, i.e. there is no possibility in U where Alice isn't RGC or where Bob isn't BYC.

What if Bob ignores the nature of Alice's colour-blindness? Suppose he knows that Alice is colour-blind but doesn't know whether Alice is RGC or BYC, assuming, for simplicity's sake, that these are the only open possibilities. What modifications must be brought to U to accommodate such assumptions? I claim that the simplest way of doing this, in the context of Kripke semantics, is to resort to a copying strategy of sorts. This strategy dictates that we give ourselves two copies V_1 and V_2 of W, one for each colour-blindness profile of Alice. We then make Alice RGC on the first copy and BYC on the second. On both copies, we make Bob BYC. Finally, we make Bob's accessibility relation join corresponding copies in V_1 and V_2 to reflect his higher-order ignorance of Alice's colour-blindness. The resulting model is represented in Fig. 3.

Alice's epistemic accessibility relation is neither RGC nor BYC, it is both. Variation in the modal facts, as was the case in the Mary example, is accomplished not by multiplying the possible accessibility relations but by multiplying the world copies.

---

[4] The reader can assume that the agents are immaterial and only have epistemic properties.

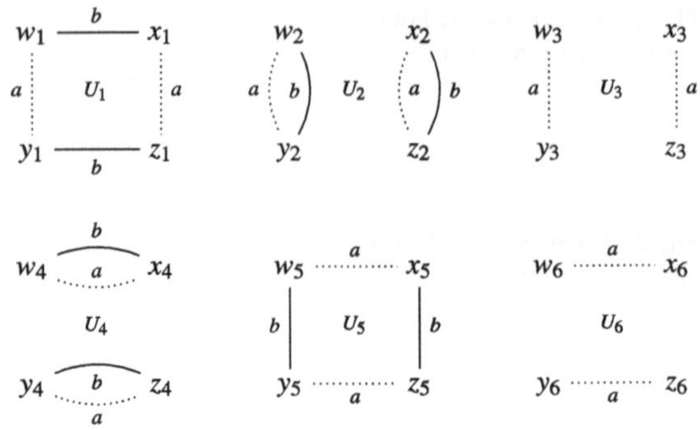

Fig. 4 Knowledge with more higher-order ignorance: part I

One may think that the added complexity of these models is acceptable, that it is a fair price to pay for the *prima facie* conceptual simplicity of possible worlds. I would like to demonstrate how high a price this is by showing how bad the copying strategy scales when we add some more second-order assumptions to the situation.

Suppose we add to the preceding scenario second-order ignorance on Alice's part. Assume, as far as Alice knows, that Bob could be RGC, BYC or not colour-blind at all (NC), and assume Bob has the same ignorance as before regarding Alice's colour-blindness. Obviously, the model defined in the previous paragraph will no longer do, because there is no possibility where Alice does not know that Bob is BYC (whichever way we choose to represent this statement in the syntax). The copying strategy dictates that we create a copy of W for each colour-blindness possible combination between Alice and Bob: copy U_1 will correspond to the combination (RGC, BYC) in which Alice is RGC and Bob is BYC, U_2 will correspond to (RGC, RGC), U_3 to (RGC, NC), U_4 to (BYC, BYC), U_5 to (BYC, RGC) and U_6 to (BYC, NC). On these various copies, we define the relations corresponding to Alice's and Bob's colour-blindness. The result is illustrated in Fig. 4.

We must then take care of connecting the worlds in different copies so as to reflect Alice's and Bob's higher-order ignorance/knowledge. Alice's epistemic accessibility relation on U (the union of the copies U_i) must connect corresponding world copies in U_1, U_2 and U_3, and it must do the same for corresponding copies in U_4, U_5 and U_6. This will reflect Alice's ignorance of Bob's specific colour-blindness. Furthermore, Bob's epistemic accessibility relation on U must connect same-copy worlds in U_1 and U_4, as well as same-copy worlds in U_2, U_5 and U_3, U_6 respectively. The result is illustrated in Fig. 5.

The intended take home message of all these examples is that conventional Kripkean semantics are ill-suited for the task of providing complex modal statements with appropriate truth-conditions, where by "appropriate" I mean not only truth-conditions that get the answer right but that do so in a principled and transparent manner. Though it may be possible to satisfy higher-order modal statements in models using modally flat worlds, it is done at the price of simplicity.

Fig. 5 Knowledge with more higher-order ignorance: part II

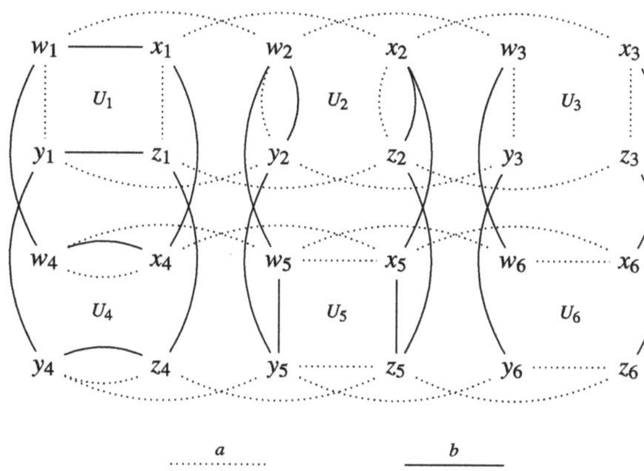

4 Possibility Spaces

An appropriate semantics for such modal combinations, we will claim, must be couched in terms of possibility spaces. A possibility space is based on the following ideas: (i) the set of possibilities (or possible worlds) is a product space; (ii) each set in this product is the set of (atomic) states of a certain type; (iii) some states, namely *modal* states, determine modal properties on other states.

If we disregard (iii), we have what is basically some form of multidimensional Kripke frame. Segerberg [10] first introduced this idea in the two-dimensional case, which constitutes the formal backbone of the notion of two-dimensional semantics as defended by Davis and Humberstone [2], and which has gained a considerable following to this day (cf. [6]). But multidimensional Kripke frames need not be exclusively thought of as devices for the expression of the contingent a priori or meta-semantic notions. Quite independently of two-dimensional semantics, Gabbay and Shehtman [3–5] use multidimensional Kripke frames to provide a semantics for what they call "products" of modal languages, basically just a language combining two or more modal languages but in a such a way that the modal notions don't "interact" with one another.[5]

In its simplest form, a multidimensional Kripke frame is a structure $\langle \mathbf{W}, \mathbf{R} \rangle$, where $\mathbf{W} = W_1 \times \cdots \times W_n$ and $\mathbf{R} = \{R_k : 1 \leq k \leq n\}$ is such that R_k is a binary relation on W_k, for each k. A language L comprising the modalities $\Diamond_k$, for each k, would be interpreted in the product structure as follows: $(\ldots, w_k, \ldots) \Vdash \Diamond_k \phi$ iff $(\ldots, v, \ldots) \Vdash \phi$, for some $v \in W_k$ such that $R_k(w_k, v)$. Basically, the k-th modality

[5] These product frames are just a special case of what Gabbay and Shehtman [3] call "fibered semantics", the general technique of mending two or more structures together to yield a further structure. Not all such combined structures have the property that the modalities have no interactions amongst themselves, thus some combined structures may actually turn out to have non-flat worlds. In fact, there is even reason to suspect that possibility spaces can be described as a special case of fibered semantics. However, since fibered semantics of this kind haven't been explored to my knowledge, thinking in those terms will not be especially useful.

moves around the k-th coordinate via the k-th relation while the other coordinate values stay fixed. However, despite being n-tuples of coordinates, these worlds are nonetheless flat in the sense we've defined, so they aren't quite what we're looking for.

Adding (iii) to the picture is what makes possibility spaces stand apart conceptually from these multidimensional Kripke frames. A modal state type is more than just another axis in the product space, the points of this axis determine properties on a subspace of the possibility space.

4.1 Formal Definition

In its most general form, a *possibility space* $\mathfrak{S}$ is defined as a triplet $\langle \tau, \Delta, \Phi \rangle$, where $\tau = \langle N, M, \mu \rangle$ is the type of $\mathfrak{S}$, Δ the domain function, and Φ the modal assignment function.

The type τ specifies the general structure $\mathfrak{S}$ has. N is the set of basic or atomic state types. Each state type $n \in N$ is a place holder for some basic or atomic feature of the underlying universe, an "axis" of the whole space, as it were. $M \subset N$ is the set of modal state types. Each modal state type pertains to a set of state types, which is specified by the function $\mu : M \rightarrow \wp(N)$.

Δ and Φ put meat on the bones of the skeleton τ describes. Δ assigns a domain of values W_n to each state type $n \in N$. The domain of the entire universe is the product space $W = \prod_{n \in N} W_n$. If $I \subset N$ is a subset of state types, then the sub-product generated by I is defined as $W_I = \prod_{n \in I} W_n$. Φ assigns an accessibility relation of the appropriate kind to each modal state. That is, if $m \in M$ is a modal state type and $w \in W_m$, then $\Phi(w)$ is an accessibility relation on $W_{\mu(m)}$.

4.2 Syntax and Semantics

We turn now to the definition of a propositional language L for possibly spaces. L is given by a set Mod of modalities and a set Prop of propositional variables (or atomic sentences). In order to interpret this language in a possibility space, some additional information is required concerning the modalities. Since modalities are in a sense the syntactic counterparts to modal features, and since there are typically many modal features in a possibility space, the model will have to specify which ones correspond to any given modality. Taking this into consideration, we define a model of L as a possibility space $\mathfrak{S} = \langle \tau, \Delta, \Phi \rangle$, where $\tau = \langle N, M, \mu \rangle$, together with an interpretation function $\mathscr{I}$ such that:

$$\mathscr{I}(\Diamond) \subset M, \text{ for all } \Diamond \in \text{Mod}$$
$$\mathscr{I}(p) \subset W, \text{ for all } p \in \text{Prop}$$

The modal state types in $\mathscr{I}(\Diamond)$ are the ones that will matter for the meaning of $\Diamond$.

In order to elegantly present the semantics, we must adopt a few abbreviations. Let $\overline{w} \in W$ and $I \subset N$. We define $\overline{w}_I$ as the restriction of $\overline{w}$ to I, i.e. $\overline{w}_I = \overline{w}\restriction_I$, and define w_n as the value $\overline{w}$ assigns to state type n. Furthermore, if $\overline{v} \in W_I$, the point $(\overline{w}_{-I}, \overline{v})$ is the element of W that is identical to $\overline{w}$ on $N \backslash I$ and identical to $\overline{v}$ on I.

Given a space $\mathfrak{S}$ and interpretation function $\mathscr{I}$, truth at a world $w \in W$ is defined recursively as follows. The propositional and Boolean clauses should be straightforward. The semantic clause for modalities goes like this: $w \Vdash \Diamond\phi$ iff, for all $m \in \mathscr{I}(\Diamond)$, we have

$$(\overline{w}_{-\mu(m)}, \overline{v}) \Vdash \phi, \text{ for some } \overline{v} \in W_{\mu(m)} \text{ such that } \Phi(w_m)(\overline{w}_{\mu(m)}, \overline{v})$$

In other words, $\Diamond$ is interpreted at $\overline{w}$ with the relation "aggregating" all the relations determined by the modal states w_m at $\overline{w}$, with $m \in \mathscr{I}(\Diamond)$. (A crucial observation to make here is that the meaning of $\Diamond$ at $\overline{w}$ is determined by the possibility $\overline{w}$ itself.) The definition of validity in a model, in a possibility space and in a collection of possibility spaces follows immediately.

5 Back to Alice and Bob

We apply the preceding to Alice and Bob's predicament. The possibility space we will use is based on the set $N = \{n_1, n_2, \ldots, n_6\}$ of basic state types, where:

n_1 = "colour of circle", with domain $\Delta(n_1) = W_1 = \{\texttt{r}, \texttt{g}\}$
n_2 = "colour of square", with $W_2 = \{\texttt{b}, \texttt{y}\}$
n_3 = "Alice's colour-blindness", with $W_3 = \{\texttt{RGC}, \texttt{BYC}\}$
n_4 = "Bob's colour-blindness", with $W_4 = \{\texttt{RGC}, \texttt{BYC}, \texttt{NC}\}$
n_5 = "Alice's higher-order ignorance", with $W_5 = \{\texttt{HI}_a\}$
n_6 = "Bob's higher-order ignorance", with $W_6 = \{\texttt{HI}_b\}$

The domains of n_5 and n_6 are singletons because, in the example above, Alice and Bob are perfectly knowledgeable about each other's higher-order ignorance. The space of possibilities W therefore looks like

$$W = \begin{Bmatrix} \texttt{r} \\ \texttt{g} \end{Bmatrix} \times \begin{Bmatrix} \texttt{b} \\ \texttt{y} \end{Bmatrix} \times \begin{Bmatrix} \texttt{RGC} \\ \texttt{BYC} \end{Bmatrix} \times \begin{Bmatrix} \texttt{RGC} \\ \texttt{BYC} \\ \texttt{NC} \end{Bmatrix} \times \{\texttt{HI}_a\} \times \{\texttt{HI}_b\}$$

The set M of modal state types of this space consists of n_3 through n_6, and it should be clear that the modal state types n_3 and n_4 both modally pertain to $\{n_1, n_2\}$, and the modal state types n_5 and n_6 both modally pertain to $\{n_3, n_4\}$.

Each of the pairs of values of $W_3 \times W_4$ defines a pair of accessibility relations on $W_1 \times W_2$, one for each Alice and Bob, as illustrated in Fig. 6. Similarly, $(\texttt{HI}_a, \texttt{HI}_b)$ determines the pair of accessibility relations illustrated in Fig. 7.

We let Φ be the function that assigns these accessibility relations. The result is the possibility space $\mathfrak{S} = \langle \tau, \Delta, \Phi \rangle$.

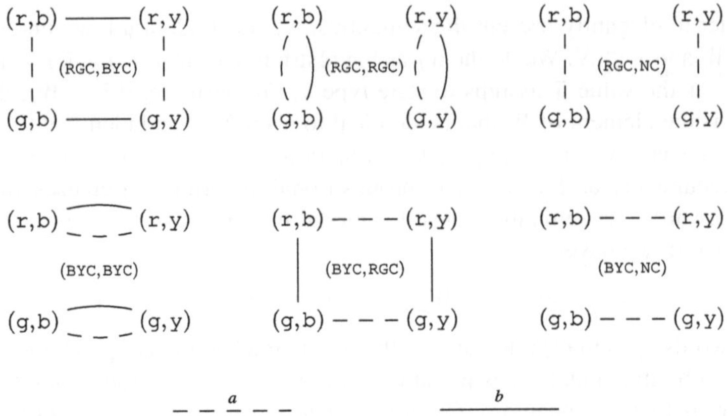

Fig. 6 Kripke frames determined by the possibility space: part I

Fig. 7 Kripke frames determined by the possibility space: part II

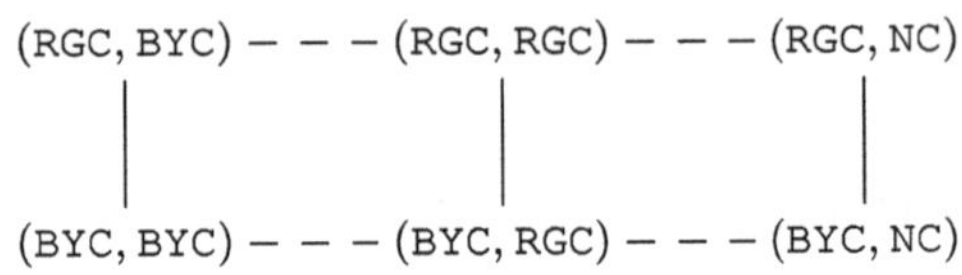

At first glance, Figs. 4 and 6 look the same (modulo renaming), but there is a crucial difference between them: only one frame in Fig. 4 is depicted whereas Fig. 6 has six of them, all based on the same set of possibilities $W_1 \times W_2$. Furthermore, Alice and Bob's higher-order ignorance is directly represented in Fig. 7, whereas in Fig. 5 it is accomplished only in a indirect way (via the curvy lines joining the world copies).

We assume the language L contains at least an epistemic modality for each Alice and Bob, say K_a and K_b. Let $\mathscr{I}$ be the interpretation implicit in this model. We have that $\mathscr{I}(K_a) = \{n_3, n_5\}$ and $\mathscr{I}(K_b) = \{n_4, n_6\}$. Let $\overline{w} \in W_1 \times \cdots \times W_6$ be a possibility. Following the modal semantic clause given above, we have that $\overline{w} \Vdash K_a\phi$ iff

$$(\overline{w}_{-\{1,2\}}, \overline{v}) \Vdash \phi, \text{ for all } \overline{v} \in W_{\{1,2\}} \text{ such that } \Phi(w_3)(\overline{w}_{\{1,2\}}, \overline{v})$$
$$(\overline{w}_{-\{3,4\}}, \overline{v}) \Vdash \phi, \text{ for all } \overline{v} \in W_{\{3,4\}} \text{ such that } \Phi(w_5)(\overline{w}_{\{3,4\}}, \overline{v}).$$

(We abbreviate the composite state types $\{n_1, n_2\}$ and $\{n_3, n_4\}$ by $\{1, 2\}$ and $\{3, 4\}$ respectively.). Similarly, we have that $w \Vdash K_b\phi$ iff.

$$(\overline{w}_{-\{1,2\}}, \overline{v}) \Vdash \phi, \text{ for all } \overline{v} \in W_{\{1,2\}} \text{ such that } \Phi(w_4)(\overline{w}_{\{1,2\}}, \overline{v})$$
$$(\overline{w}_{-\{3,4\}}, \overline{v}) \Vdash \phi, \text{ for all } \overline{v} \in W_{\{3,4\}} \text{ such that } \Phi(w_6)(\overline{w}_{\{3,4\}}, \overline{v}).$$

These truth conditions for K_a and K_b highlight the way in which the various parts of the world come together to define the agents' knowledge.

6 Concluding Remarks

I have defended the claim that possible worlds should be understood as structured entities with modal parts. Possible worlds with modal parts allow for modal variation in the semantics, and this variation leads to a more natural representation of the truth conditions of complex modal statements. In particular, we have seen how epistemic modalities (and other modal combinations) are well served by such a framework. Applications are obviously not limited to that realm, however. One unsuspected application of the system is to counterfactual conditionals. It can be shown that systems of spheres (cf. [7]) are a special case of possibility space. Since we can easily graft other modal features to any possibility space, the upshot is that this provides us with a general semantics for tensed counterfactuals or counterfactual knowledge.

It should also be mentioned that we can prove correspondence and completeness results for possibility space semantics, but this goes well beyond the scope of the present paper. To get a hold on the product structure and the modal properties, we require a highly customized version of the language of hybrid logic. The axiomatization and completeness results are highly adapted from Blackburn et al. ([1], pp. 434–445).

Acknowledgments This work was supported by grant # 149410 of the FQRSC. I would like to thank audiences at *Trends in Logic 2012* and *Logica 2012*, and I'm grateful for comments from Agustin Rayo.

References

1. Blackburn, P., de Rijke, M., & Venema, Y. (2001). *Modal logic, Cambridge tracts in theoretical computer science* (Vol. 53). Cambridge: Cambridge University Press.
2. Davies, M., & Humberstone, L. (1980). Two notions of necessity. *Philosophical Studies*, *38*, 1–30.
3. Gabbay, D., & Shehtman, V. B. (1998). Products of modal logics: part 1. *Logic Journal of IGPL*, *6*(1), 73.
4. Gabbay, D., & Shehtman, V. B. (2000). Products of modal logics: part 2. *Logic Journal of IGPL*, *8*(2), 165–210
5. Gabbay, D., & Shehtman, V. B. (2002). Products of modal logics: part 3. *Studia logica*, *72*(2), 157–183
6. García-Carpintero, M., Macià, J. (Eds.). (2006). *Two-dimensional semantics*. Oxford: Oxford University Press.
7. Lewis, D. K. (1973). Counterfactuals. Oxford: Blackwell.
8. Lewis, D. K. (1986). *On the plurality of worlds*. Oxford: Blackwell. NULL
9. Prior, A. (1967). *Past, present and future*. Oxford: Oxford University Press.
10. Segerberg, K. (1973). Two-dimensional modal logic. *Journal of Philosophical Logic*, *2*(1), 77–96.

Collective Alternatives

Franz von Kutschera

Abstract In the logic of agency individual alternatives of agents have been taken as basic, so far, while the alternatives of groups of agents have been derived from the alternatives of the group members. In many cases, however, groups have additional possibilities. In the paper I propose a generalized theory of collective alternatives that takes them as fundamental.

Keywords Collective agency · Collective alternatives · Logic of agency · T × W frames

A logic of actions and of bringing about something by an action is useful in many realms of philosophy, from moral philosophy to the philosophy of mind. The logic of agency has been developed mainly by Lennart Åqvist [1, 2], Franz von Kutschera [3, 4] and Nuel Belnap [5, 6]. In this logic individual alternatives of agents have been taken as basic, so far, while the alternatives of groups of agents were derived from those of the group members. In many applications this approach turns out to be insufficient. Often groups have possibilities that are not combinations of the separate possibilities of their members. A generalization of the logic of agency therefore is called for, and that is the topic of my paper.

It is concerned only with the semantics of a logic of agency, not with its language or axiomatics. The appropriate semantical framework is that of branching worlds or world histories, especially T × W-frames with a universal time order for all histories. I shall first present this framework and define individual and collective alternatives in the usual way. I will then give two examples for the failure of this definition of collective alternatives. They suggest a definition of alternatives that takes collective ones as fundamental and permits the individuals to have alternatives in a group that they dont have by themselves. A further example will then show that the collective

F. von Kutschera (✉)
University of Regensburg, Regensburg, Germany
e-mail: franz.kutschera@psk.uni-regensburg.de

R. Ciuni et al. (eds.), *Recent Trends in Philosophical Logic*,
Trends in Logic 41, DOI: 10.1007/978-3-319-06080-4_10,

alternatives of a group have to be defined relatively to coalitions among the other agents.

1 T x W Frames and Tensed Possibilities

I assume knowledge of T × W frames and just present the definition.

(**D1**) A $T \times W$ frame is a quadruple $U = (T, <, W, \sim)$ such that:

(a) T is a non-empty set of moments of time $(t, t', \ldots)$.
(b) $<$ is a linear ordering on T; $t < t'$ means that t is earlier than t'.
(c) W is a non-empty set of worlds $(w, w', \ldots)$.
(d) For all $t \in T$, $w \sim_t w'$ is an equivalence relation on W, read as *w coincides with w′ in t*, such that $w \sim_t w' \wedge t' \leq t \rightarrow w \sim_{t'} w'$—worlds divide only into the future.
(e) $w \neq w' \rightarrow \exists t \forall t'(t' \leq t \leftrightarrow w \sim_{t'} w')$—different worlds alsways have a last point of coincidence.

Condition (e) corresponds to the intiuition that different choices make a difference only for the future.

I shall use two abbreviations:

(**D2**) (a) $P(t, w) := \{w' : w' \sim_t w\}$—the set of worlds possible in w at t.
(b) $C(t, w) := \{w' : \exists t'(t < t' \wedge w' \sim_t w)\}$—the set of w and the worlds departing from w only in the future.

In T × W frames states of affairs obtain in worlds at times. Therefore a state of affairs can be represented by the set of pairs of times and worlds, in which it holds. If and only if for one of these worlds w' in $P(t, w)$ the pair (t, w') belongs to the state of affairs A, A is possible in w and t. This possibility does not just depend on worlds but also on times.

2 Individual Alternatives

An *action* of a person is a behaviour she can also refrain from. An action always starts from a situation in which the agent has at least two alternatives. If we describe the behaviour of a person as an "action" we always presuppose that she could do otherwise. That is not always the case if it is possible that she behaves differently. If somebody falls down the stairs, for instance, that is normally a contingent event. Therefore it was possible that he would not fall. But this does not make his fall an action. In the case of an action it has to be *possible for the agent* to do otherwise and this possibility must consist in an alternative that was open to him. We have to distinguish therefore between *event possibility* as it was just defined, possibility, e.g. by the laws of nature, and *agent possibility*, possibility for an agent.[1]

To represent actions we start from a T×W frame U and add a finite set $G = \{g_1, \ldots, g_n\}$ of agents. In every world w and at each moment t every agent g has a set $A(g, t, w)$ of alternatives.

(D3) A *system of individual alternatives* based on U is a pair (G, A) such that:

(a) G is a set of agents, $g_1, \ldots, g_n$.
(b) For all $g \in G, w, t$: $A(g, t, w)$ is the set of alternatives of the agent g in w and t. These sets have the following properties:

(b1) $w' \in P(t, w) \rightarrow A(g, t, w) = A(g, t, w')$.
(b2) $X \in A(g, t, w) \rightarrow \emptyset \neq X \subseteq P(t, w)$.
(b3) $w' \in X \wedge X \in A(g, t, w) \rightarrow C(t, w') \subseteq X$.
(b4) $X, Y \in A(g, t, w) \rightarrow X = Y \vee X \cap Y = \emptyset$.
(b5) $P(t, w) \subseteq \bigcup A(g, t, w)$.
(b6) $X_1 \in A(g_1, t, w) \wedge \ldots \wedge X_n \in A(g_n, t, w) \rightarrow X_1 \cap \ldots \cap X_n \neq \emptyset$.

Comments: (b1) Alternatives do not depend on the future. (b2–5) The sets of individual alternatives in w at t are divisions (partitions) of the set $P(t, w)$ of possible worlds. (b3) Agents cannot discriminate between worlds that branch only at a later moment. (b6) No alternative can be blocked by choices of the other agents.

Not every agent has a choice at every moment. Therefore sets of alternatives $A(g, t, w)$ are admitted containing $P(t, w)$ as the only alternative. In this case I shall say that g has no genuine alternative, no alternative he could refrain from realizing. An agent has a genuine alternative only if he has at least two alternatives.

It is often useful to add the condition

(b7) $w' \in P(t, w) \rightarrow \exists X_1 \ldots X_n(X_1 \in A(g_1, t, w) \wedge \ldots \wedge X_n \in A(g_n, t, w) \wedge X_1 \cap \ldots \cap X_n = C(t, w'))$.

This is a completeness condition: The agents in G together can determine how the world goes on after t. Generally this condition is tenable only if we count Mother Nature, which is responsible for chance events, among the agents.

With respect to these alternatives we can define agent possibility: *For the agent* g it is *possible* in w at t to bring about the state of affairs A, if g has a genuine alternative in w, t for which A holds at t in all the worlds of this alternative. An agent g *brings it about* in w at t that A holds at t, if the alternative g realizes in w at t is a subset of the set of all worlds in which A holds at t.

[1] On the relation of the statements "The agent X has the possibility to do F" and "It is possible, that X does F", and the relation between "X could have done otherwise" and "If X would have wished differently he would have acted differently" there is a whole library of publications. Determinists naturally misinterpret the first sentences in the sense of the latter. Cf. e.g. ([8], Chap. 6).

3 Collective Alternatives: The Usual Approach

Alternatives of groups of agents from G, are normally defined by individual alternatives: If $A(\{g_{i1}, \ldots, g_{im}\}, t, w)$ is the set of alternatives of the group $\{g_{i1}, \ldots, g_{im}\}$ in w, t, we have:

(D4) $A(\{g_{i1}, \ldots, g_{im}\}, t, w) :=$
$\{X_1 \cap \ldots \cap X_m : X_1 \in A(g_{i1}, t, w) \wedge \ldots \wedge X_m \in A(g_{im}, t, w)\}$

The alternatives of a group, therefore, are the combinations of the individual alternatives of its members.

We have for $G', G'' \subseteq G$

(a) $X \in A(G', t, w) \wedge Y \in A(G'', t, w) \wedge G' \cap G'' = \emptyset \rightarrow X \cap Y \in A(G' \cup G'', t, w)$.
(b) $X \in A(G', t, w) \wedge G' \subseteq G'' \rightarrow \exists Y(Y \in A(G'', t, w) \wedge Y \subseteq X)$.

4 Counterexamples

According to **D4** the alternatives of a group result from the alternatives of its members. In realizing a collective alternative they do in coordination, what they can also do separately. There are, however, many cases in which groups have new possibilities, possibilities beyond those envisaged by **D4**. The following two examples show that co-operation opens up new possibilities of action.

Case 1: Peak A
Two mountaineers can either climb peak B separately, a lower pinnacle in front of peak A, or they can climb A together, as a team. Each of them has two individual alternatives to stay in the camp or to climb peak B—but together they have the additional alternative of climbing A as a team. This alternative does not arise from the separate possibilities in the way stated in **D4**.

Case 2: The Ruritarian Prison Cell
In Ruritania prison cells for two occupants are so small that there is only room for one person to sit while the other has to stand. The occupants of such a cell have no genuine individual alternative. They cannot sit or stand independently of what the other does, so that, without cooperation, their positions will have to remain as they are. Only in a coalition they have genuine alternatives and can determine, who sits and who stands. These collective alternatives again do not result from individual ones.

5 Collective Alternatives: A New Approach

These examples suggest that we conceive of collective alternatives not as combinations of individual alternatives as in **D4** but as fundamental. We cannot, however, define alternatives of the type $A(G', t, w)$ where G' is a subset of G, the set of all the agents. Rather the alternatives of groups are sensitive to what co-operations are possible between the other members of G. This is shown by the following example:

Case 3: The last glass of rum
John, Tom and Max each want to have what is left in a bottle of rum. John is stronger than each of the other two but together they can hold him back. So if Tom and Max cooperate John has no alternative, but if there is no co-operation between Tom and Max, John may drink the rest of the rum or leave it to the others, as he pleases. His alternatives depend on what co- operations are possible between the others.

We therefore have to define collective alternatives relatively to admissible co-operations or coalitions among the rest of the agents. Coalitions are defined by partitions $D = \{G_1, \ldots, G_m\}$ of the set G of agents. So we consider sets of alternatives $A(G_i, D, t, w)$ for $G_i \in D$. For the individual alternatives envisaged in **D3** we have

$$A(g, t, w) = A(\{g\}, D_0, t, w) \text{ for } D_0 = \{\{g_1\}, \ldots, \{g_n\}\}$$

and the collective alternatives of **D4**

$$A(\{g_{i1}, \ldots, g_{im}\}, t, w) = A(\{g_{i1}, \ldots, g_{im}\}, (D_0 - \{\{g_{i1}\}, \ldots, \{g_{im}\}\}) \cup \{g_{i1}, \ldots, g_{im}\}, t, w).$$

If we consider the groups $G_1, \ldots, G_m$ in a partition D of G as individuals we get conditions corresponding to those of **D3**. The main difference is that the alternatives in $A(G_i, D, t, w)$ are partitions not of $P(t, w)$, the set of all possible worlds, but of a non-empty subset $P(D, t, w)$ of $P(t, w)$, the set of possible outcomes for coalition structure D.

(**D5**) A *system of collective alternatives* based on a tree-universe U is pair (G, A) such that:

(a) G is a set of agents, $g_1, \ldots, g_n$.
(b) For all partitions $D = \{G_1, \ldots, G_m\}$ of G and all w, t: $A(G_i, D, t, w)$ is the set of alternatives of the group G_i relative to the partition D in w and t. For $P(D, t, w) := \bigcup_{1 \leq i \leq m} A(G_i, D, t, w)$, these sets have the following properties:
  (b1) $P(D, t, w) \subseteq P(t, w)$.
  (b2) $w' \in P(t, w) \rightarrow A(G_i, D, t, w) = A(G_i, D, t, w')$.
  (b3) $X \in A(G_i, D, t, w) \rightarrow \emptyset \neq X \subseteq P(D, w, t)$.
  (b4) $w' \in X \wedge X \in A(G_i, D, t, w) \rightarrow C(t, w') \subseteq X$.
  (b5) $X, Y \in A(G_i, D, t, w) \rightarrow X = Y \vee X \cap Y = \emptyset$.
  (b6) $P(D, w, t) \subseteq \bigcup_{1 \leq i \leq m} A(G_i, D, t, w)$.
  (b7) $X_1 \in A(G_1, D, t, w) \wedge \ldots \wedge X_m \in A(G_m, D, w, t) \rightarrow X_1 \cap \ldots \cap X_m \neq \emptyset$.
  (b8) $X \in A(G_i, D, w, t) \wedge Y \in A(G_k, D, w, t) \rightarrow \exists Z(Z \in A(G_i \cup G_k, D - \{G_i, G_k\}) \cup \{G_i \cup G_k\}, w, t) \wedge Z \subseteq X \cap Y)$.

Comment: (b2–7) are taken over from **D3**. (b8) corresponds to **D4**; *case 2* shows that $Z \subseteq X \cap Y$ cannot be replaced by $Z = X \cap Y$. From (b8) we get

(c) $X \in A(G_i, D, t, w) \to \exists Y(Y \in A(G, \{G\}, t, w) \wedge Y \subseteq X)$

the biggest coalition G can bring about everything that smaller coalitions can bring about.

The completeness condition corresponding to **D3**, (b7) is:

(b9) $w' \in P(t, w) \to C(t, w) \in A(G, \{G\}, t, w)$.

If we count Mother Nature, n, among the agents we should only consider coalition structure D such that $\{n\} \in D$, since there can be no cooperation with chance.

From (b9) we obtain $P(t, w) \subseteq P(\{G\}, t, w)$, and in view of (b1)

(d) $P(t, w) = P(\{G\}, t, w)$ and
(e) $P(D, t, w) \subseteq P(\{G\}, t, w)$.

Collective alternatives are more general than dependent alternatives obtained from **D3** by dropping (b6), as our case 1 shows.[2]

References

1. Åquist, L. (1974). A new approach to the logical theory of action and causlity. In Stenlund S (Ed.), *Logical theory and semantic analysis*, (pp. 73–91). Dordrecht: Reidel.
2. Åquist, L., Mullock, P. (1989). *Causing harm: A logico-legal study*. Berlin: Walter de Gruyter.
3. Belnap, N. (1991). Before refraining. *Erkenntnis*, *34*(2), 137–169.
4. Belnap, N., & Perloff, M. (1988). Seeing to it that: A canonical form of agentives. *Theoria*, *54*(3), 175–199.
5. Kutschera, F. (1986). Bewirken. *Erkenntnis*, *24*(3), 253–281.
6. Kutschera, F. (1993). Causation. *Journal of Philosophical Logic*, *22*(6), 563–588.
7. Kutschera, F. (2011). Individuelle und kollektive Alternativen. In C. Lumer & U. Meyer (Eds.), *Geist und Moral* (pp. 279–295). Mentis: Analytische Reflexionen für Wolfgang Lenzen.
8. Smart, J. J. (1966). *Philosophy and scientific realism*. London: Routledge and Paul.

[2] Relations to the treatment of collective agency in other logics are stated in [7].

da Costa Meets Belnap and Nelson

Hitoshi Omori and Katsuhiko Sano

Abstract There are various approaches to develop a system of paraconsistent logic, and those we focus on in this paper are approaches of da Costa, Belnap, and Nelson. Our main focus is da Costa, and we deal with a system that reflects the idea of da Costa. We understand that the main idea of da Costa is to make explicit, within the system, the area in which you can infer classically. The aim of the paper is threefold. First, we introduce and present some results on a classicality operator which generalizes the consistency operator of Logics of Formal Inconsistency. Second, we show that we can introduce the classicality operator to the systems of Belnap. Third, we demonstrate that we can generalize the classicality operator above to the system of Nelson. The paper presents both the proof theory and semantics for the systems to be introduced, and also establishes some completeness theorems.

Keywords Paraconsistent logic · Paracomplete logic · Four-valued logic · Consistency operator · Classicality operator

1 Introduction

"The notion of a theory's being trivial must be distinguished from its being contradictory." This is the slogan for paraconsistent logic. Beyond this, paraconsistent logicians disagree on many points. Consequently, many different approaches to systems of paraconsistent logic have been developed.

H. Omori (✉)
The Graduate Center, CUNY, 365 Fifth Avenue, New York, NY 10013, USA
e-mail: hitoshiomori@gmail.com

H. Omori
Department of Philosophy II, RUB, Universitätsstraße 150, D-44780 Bochum, Germany

K. Sano
School of Information Science, Japan Advanced Institute of Science and Technology, 1-1 Asahidai, Nomi, Ishikawa 923-1292, Japan
e-mail: v-sano@jaist.ac.jp

R. Ciuni et al. (eds.), *Recent Trends in Philosophical Logic*,
Trends in Logic 41, DOI: 10.1007/978-3-319-06080-4_11,

Table 1 List of four-valued truth tables for logical connectives of **BS4**

A	$\sim A$	$\circ A$
t	**f**	**t**
b	**b**	**f**
n	**n**	**f**
f	**t**	**t**

$A \wedge B$	**t**	**b**	**n**	**f**
t	**t**	**b**	**n**	**f**
b	**b**	**b**	**f**	**f**
n	**n**	**f**	**n**	**f**
f	**f**	**f**	**f**	**f**

$A \vee B$	**t**	**b**	**n**	**f**
t	**t**	**t**	**t**	**t**
b	**t**	**b**	**t**	**b**
n	**t**	**t**	**n**	**n**
f	**t**	**b**	**n**	**f**

$A \supset B$	**t**	**b**	**n**	**f**
t	**t**	**b**	**n**	**f**
b	**t**	**b**	**n**	**f**
n	**t**	**t**	**t**	**t**
f	**t**	**t**	**t**	**t**

The main subject of this paper is the notion of consistency operator which reflects the idea of da Costa, and later introduced and studied intensively by Carnielli, Coniglio, Marcos and their collaborators by developing a family of systems known as Logics of Formal Inconsistency (LFIs hereafter). The idea of da Costa was to control the behavior of contradictions by means of the notion of consistency operator so that contradictions do not always explode. There are infinitely many systems of paraconsistent logic which reflect this idea, and many criticisms against those systems are known. We may raise two of them. One is that its semantics is non-compositional and is therefore difficult to grasp, and the other is that material conditional, which sometimes makes the system trivial together with other non-logical axioms, is present.

Before turning to these objections, let us clarify our understanding of da Costa's idea. One of the features of LFIs is that Law of Excluded Middle (LEM) with respect to paraconsistent negation is always assumed in those systems. This is probably influenced by one of the four criterions for paraconsistency given by da Costa (cf. [6, p. 498]) in which he requires paraconsistent systems to "contain the most part of the schemata and rules of $\mathbf{C_0}$" where $\mathbf{C_0}$ is the classical propositional calculus.[1] In view of this criteria, it seems to be reasonable to assume LEM with respect to paraconsistent negation since in many cases, though not always, validity of LEM is independent of validity of *ex contradictione quodlibet*. But at the same time, it has been realized especially by the system studied by Belnap and Dunn, that not only inconsistency, but also incompleteness should be taken care of in certain situations. In these cases, da Costa's idea can be understood as follows: make explicit the cases when we can apply inferences of classical logic. And if we accept this understanding, it is more appropriate to refer to the characteristic connective reflecting da Costa's idea not as consistency but as classicality or normality. In this paper, we make use of the former.

With these observations in mind, let us now consider the above two objections. For this purpose, we start with a four-valued system **BS4** introduced in [13] along the lines of research of LFIs. **BS4** is proved to be complete with respect to the following truth tables (cf. Theorem 2) (Table 1):

Note here that the designated values are **t** and **b** (we denote by $\mathcal{D}$ the set of designated values). In view of our understanding of da Costa's idea, the above truth table for $\circ$, namely the classicality operator, must be reasonable. Indeed, it clearly distinguishes

[1] Interestingly, a similar criteria is also considered by Jaśkowski (cf. [8, p. 38]) who is the other founder of modern paraconsistent logic.

(A1) $\circ A \supset (A \vee \sim A)$	(A5) $\sim(A \vee B) \equiv (\sim A \wedge \sim B)$
(A2) $\circ A \supset (A \supset (\sim A \supset B))$	(A6) $\sim(A \supset B) \equiv (A \wedge \sim B)$
(A3) $(A \supset \circ A) \equiv (\sim A \supset \circ A)$	(A7) $\sim\sim A \equiv A$
(A4) $\sim(A \wedge B) \equiv (\sim A \vee \sim B)$	(A8) $\sim(\circ A) \equiv (A \equiv \sim A)$

'classical' values **t** and **f** from 'non-classical' values **b** and **n**. It is also possible to consider a consistency operator as well. We shall consider this in Sect. 3.

Now, since the four-valued tables for conjunction, disjunction and negation coincide with that of Belnap [1], this must settle, at least to a certain extent, the first objection that the semantics for systems following da Costa's idea are non-compositional.[2]

However, this four-valued system still keeps the material conditional. This is because the approach taken by da Costa and his followers accept the validity of material conditional. But the challenge provided by the second objection is important for the understanding of the notion of classicality operator. So the main question in this paper is: Can the notion of classicality operator be introduced in systems without the help of material conditional? This question may be interpreted at least in the following two ways:

- Can the notion of classicality operator be introduced in systems in which the material conditional is not definable?
- Can the notion of classicality operator be generalized to systems in which a conditional different from material conditional is definable?

In the present paper, we will consider the following two questions which are special cases of the above two questions.

- Can the notion of classicality operator be introduced in four-valued system of Belnap and Dunn without material conditional being definable?
- Can the notion of classicality be generalized to Nelson's system within which a constructive conditional is deployed instead of the material conditional?

Note here that there is still a worry on what we mean by classicality operator, since the semantic frameworks for Belnap-Dunn logic and Nelson logic are not necessarily the same. Our original motivation was to begin with the system **BS4**, and shed some light on the classicality operator introduced in that system. Therefore, to be more precise, the question we deal with should be read as follows.

- Can the classicality operator of **BS4** be introduced in a four-valued system of Belnap and Dunn without the material conditional being definable?
- Can the classicality operator of **BS4** be generalized to Nelson's system within which a constructive conditional is deployed instead of the material conditional?

---

[2] Note here that **BS4** is *not* the first many-valued system that reflects da Costa's idea. Indeed, there are systems such as **LFI1** and **LFI2**, developed in [4], which are complete with respect to three-valued semantics. Therefore, we may say that the first objection was already settled then. But at the same time, we are widening our scope to deal with incomplete situations and therefore it must
(Footnote 2 continued)
be fair to say that the argument against the first objection becomes more widely acceptable by the presence of the system **BS4**.

Then, the purpose of the present paper is to answer these two questions in the affirmative.[3] In short, our results show that the idea of classicality operator does not necessarily force us to accept material conditional. We also suggest a tentative understanding of classicality based on the results we present in this paper.

The paper is organized as follows. After some preliminaries in the next section, we review and generalize the LFIs, and present a thesis which characterizes the classicality operator of **BS4** in the third section. This will be a starting point for our discussion. We also give the full proof of the completeness theorem for **BS4** which was roughly sketched in [13]. Then, we turn to two questions we addressed above which will be taken up in sections four and five respectively. In particular, in order to answer the second question, we make use of the characteristic thesis above also in a constructive context. Finally, the sixth section concludes the paper.

2 Preliminaries

Before starting our discussion, let us provide some preliminaries. First, our syntax consists of a finite set P of propositional connectives and a countable set Var of propositional variables which we refer to as $\mathscr{L}_{\mathsf{P}}$. Furthermore, we denote by $\mathsf{Form}_{\mathsf{P}}$ the set of formulas defined as usual in $\mathscr{L}_{\mathsf{P}}$. In this paper, we always assume that $\{\sim, \wedge, \vee\} \subseteq \mathsf{P}$ and just include the propositional connective(s) different from $\{\sim, \wedge, \vee\}$ in the subscript of $\mathscr{L}_{\mathsf{P}}$. For example, we write $\mathscr{L}_{\circ}$ and $\mathsf{Form}_{\circ}$ to mean $\mathscr{L}_{\{\sim,\wedge,\vee,\circ\}}$ and $\mathsf{Form}_{\{\sim,\wedge,\vee,\circ\}}$ respectively. Moreover, we denote a formula of $\mathscr{L}_{\mathsf{P}}$ by A, B, C, etc. and a set of formulas of $\mathscr{L}_{\mathsf{P}}$ by Γ, Δ, Σ, etc.

Second, we need to specify our four-valued semantics. Given the set $\mathsf{Form}_{\mathsf{P}}$ of the formulas of $\mathscr{L}_{\mathsf{P}}$, we define the notions of valuation and semantic consequence in terms of the four-values **t**, **b**, **n**, and **f**. Intuitively speaking, a valuation is a 'homomorphism' from the 'term algebra' $(\mathsf{Form}_{\mathsf{P}}, \mathsf{P})$ to an algebra $(\{\mathbf{t}, \mathbf{b}, \mathbf{n}, \mathbf{f}\}, \mathsf{P})$.

Definition 1 Define $E := \{\mathbf{t}, \mathbf{b}, \mathbf{n}, \mathbf{f}\}$. A *four-valued valuation for* $\mathsf{Form}_{\mathsf{P}}$ is the unique extension $v : \mathsf{Form}_{\mathsf{P}} \to E$ of a mapping $\mathsf{Var} \to E$ that is induced by the truth tables for connectives of P listed in Table 1.

Definition 2 Define the set $\mathscr{D} \subseteq E$ of the *designated values* as $\{\mathbf{t}, \mathbf{b}\}$. Let $\Sigma \cup \{A\}$ be a subset of $\mathsf{Form}_{\mathsf{P}}$. Then, A is a *four-valued semantic consequence* from Σ (notation : $\Sigma \models_4 A$) if, for all four-valued valuations v for $\mathsf{Form}_{\mathsf{P}}$ such that $v(B) \in \mathscr{D}$ for all $B \in \Sigma$, $v(A) \in \mathscr{D}$. A formula $A \in \mathsf{Form}_{\mathsf{P}}$ is a *four-valued tautology* if $\emptyset \models_4 A$, i.e., $v(A) \in \mathscr{D}$ always holds for any four-valued valuation v for $\mathsf{Form}_{\mathsf{P}}$.

Third, the following is a list of axioms which we make use of in the present paper. Note here that $A \equiv B$ is defined as $(A \supset B) \wedge (B \supset A)$ as usual.

Fourth, the rules of Table 2 are the rules for the natural deduction system which we refer to in the present paper. Then, we introduce the base system of natural deduction in this paper as follows.

[3] Note that two questions are quite different and thus our approaches to these questions will be also quite different.

Table 2 Natural deduction rules of this paper

$$\frac{\sim\sim A}{A}\ (\sim\sim) \qquad \frac{\sim(A\wedge B)}{\sim A\vee\sim B}\ (\sim\wedge) \qquad \frac{\sim(A\vee B)}{\sim A\wedge\sim B}\ (\sim\vee)$$

(the three rules above are printed with double lines)

$$\frac{}{A\vee(A\supset B)}\ (\supset_1) \qquad \frac{\sim(A\supset B)}{A\wedge\sim B}\ (\supset_2)$$

(rule $(\supset_2)$ is printed with a double line)

$$\frac{\circ A \quad A \quad \sim A}{B}\ (\circ_1) \qquad \frac{\circ A}{A\vee\sim A}\ (\circ_2) \qquad \frac{\sim\circ A \quad A}{\sim A}\ (\circ_3) \qquad \frac{\sim\circ A \quad \sim A}{A}\ (\circ_4) \qquad \frac{}{\circ\circ A}\ (\circ_5)$$

$$\frac{A\supset\circ A \quad \sim A}{\circ A}\ (\circ\supset_1) \qquad \frac{\sim A\supset\circ A \quad A}{\circ A}\ (\circ\supset_2) \qquad \frac{A\supset\sim A \quad \sim A\supset A}{\sim\circ A}\ (\sim\circ\supset)$$

$$\frac{\begin{matrix}[A]\\ \vdots\\ \circ A\end{matrix} \quad \sim A}{\circ A}\ (\mathbf{c}\circ_1) \qquad \frac{\begin{matrix}[\sim A]\\ \vdots\\ \circ A\end{matrix} \quad A}{\circ A}\ (\mathbf{c}\circ_2) \qquad \frac{\begin{matrix}[\sim A]\\ \vdots\\ A\end{matrix} \quad \begin{matrix}[A]\\ \vdots\\ \sim A\end{matrix}}{\sim\circ A}\ (\mathbf{c}\sim\circ)$$

Definition 3 $\mathscr{N}$**BD** is a system of natural deduction which consists of the inference rules $(\sim\sim)$, $(\sim\wedge)$, and $(\sim\vee)$ as well as all the introduction and elimination rules of $\wedge$ and $\vee$.

Note here that although there are some attempts of developing natural deduction systems for da Costa's systems $\mathbf{C}_n$ (e.g. [5], and some of the references therein), Hilbert-style presentation has always been a preferred style to develop proof theories in LFIs. But in the present paper, we develop the proof theory not only by Hilbert-style system but also by natural deduction systems, following the presentation of [15], with a hope of adding some new perspectives on the behavior of consistency operator.

3 Logics of Formal Inconsistencies Revisited and Generalized

Following the pioneering works of da Costa and his collaborators on systems $\mathbf{C}_n$ [6], important progress was made by Carnielli, Marcos, and Coniglio [2, 3]. Their contribution is that they generalized the consistency so that the system has the notion of consistency operator as a *primitive* connective not as a *defined* connective as in da Costa's systems $\mathbf{C}_n$. This generalization enabled them to clarify the essence of da Costa's idea. However, the studies of LFIs were limited to systems which are paraconsistent but not paracomplete, and the case when the systems are both paraconsistent and paracomplete remained unexplored.[4] Based on these, we first

[4] Note here that there is an attempt [10] to develop systems which are both paraconsistent and paracomplete, following the line of research of da Costa. Recall that a system is called paracomplete when the law of excluded middle is not valid in the system.

develop a new system which is both paraconsistent and paracomplete, and then extend it to a four-valued system. Note here that we work with the language $\mathscr{L}_{\circ,\supset}$ in this section except in the considerations on consistency operator.

Definition 4 Let **CPC**$^+$ be the extension of the negation-less fragment of intuitionistic propositional calculus (**IPC**$^+$) enriched by $A \vee (A \supset B)$.[5] Then, systems **cBS** and **BS** are obtained by adding axioms (A1)–(A3) to **IPC**$^+$ and **CPC**$^+$ respectively. Furthermore, **mbC**, the base system in [2], consists of (A2) and $A \vee \sim A$ together with **CPC**$^+$.

Definition 5 A unary operation $\neg$ called *strong negation*[6] is defined as $\neg A := A \supset (\circ X \wedge X \wedge \sim X)$ for some X in the language $\mathscr{L}_{\circ,\supset}$.

Some of the differences between **mbC** and **BS** are as follows. First, the law of excluded middle with respect to $\sim$ holds without any restriction in **mbC** whereas it is restricted in **BS** as above. This makes the system **BS** not only paraconsistent, but also paracomplete. Second, (A3) does *not* hold in **mbC** whereas it does in **BS**. One of the consequences of this fact is realized in the following theorem.

Theorem 1 *The following formula, which can be regarded as a characterization of classicality, is provable in* **cBS**, *and thus in* **BS**, *but not in* **mbC**:

$$\circ A \equiv (\neg(A \wedge \sim A) \wedge (A \vee \sim A)) \tag{1}$$

where $\neg$ *is the strong negation.*

Proof An outline of the proof for (1) in **cBS** is as follows. Since the left-to-right direction is immediate by the axioms (A1) and (A2), we outline the other way around. Let us start with $(\neg(A \wedge \sim A) \wedge (A \wedge \sim A)) \supset \circ A$, a special case of *ex contradictione quodlibet* with respect to $\neg$, which is equivalent to $(\neg(A \wedge \sim A) \wedge A) \supset (\sim A \supset \circ A)$. By the right-to-left direction of (A3), we obtain $(\neg(A \wedge \sim A) \wedge A) \supset (A \supset \circ A)$, and then by the contraction of the premises, we obtain $(\neg(A \wedge \sim A) \wedge A) \supset \circ A$. Likewise, by taking the left-to-right direction of (A3), we obtain $(\neg(A \wedge \sim A) \wedge \sim A) \supset \circ A$. Recalling that having $(A \wedge B) \supset D$ and $(A \wedge C) \supset D$ is equivalent to have $(A \wedge (B \vee C)) \supset D$ in general, we thus get $(\neg(A \wedge \sim A) \wedge (A \vee \sim A)) \supset \circ A$, the desired result.

As for the non-provability of (1) in **mbC**, consider the ordinary two valued truth tables for the classical propositional calculus, and always assign the value false to

---

[5] We may of course add the Peirce's law, i.e. $((A \supset B) \supset A) \supset A$, in place of $A \vee (A \supset B)$ Indeed, these two formulas are equivalent in negation-less fragment of classical propositional calculus as $A \vee B$ and $(A \supset B) \supset B$ are equivalent in general. The reason we employed $A \vee (A \supset B)$ is because we are simply following the convention in the study of LFIs. For an example of an axiomatization of **IPC**$^+$, see e.g. (1)–(9) of [6, pp. 498–499] which is the axiomatization given by Kleene.

[6] Note here that in the study of Nelson's systems, we also use the name strong negation to refer to a different negation. But we will here follow the convention of LFIs, not of Nelson's systems, since our main focus is on the systems that generalize the framework of LFIs. Note further that strong negation $\neg$ behaves as classical negation in **BS**.

$\circ A$. Then, these truth tables validate the axioms of **mbC** and modus ponens preserves the value true, but (1) takes the value false, as desired. □

Remark 1 First, note that the two valued truth tables deployed above show that (A3) itself is not provable in **mbC**. Second, it also seems to be interesting to explore the subsystem of **BS** that can be obtained by eliminating (A3) since it will enable us to handle all the existing LFIs in a uniform perspective which is not possible by **BS**. The further details on the system **BS** and its subsystems will be kept for another occasion. Finally, although the main questions considered in this paper is classicality operator of **BS4**, it seems that an important characterization from a proof theoretical viewpoint can be given already in a weak system such as **cBS**. But again, further examination of (1) will be kept for another occasion.

Now we extend the system **BS** into a four-valued system. This kind of system had not existed and one of the reasons seems to be that the scope of LFIs was limited to paraconsistent but not paracomplete systems, as we noted above.

Definition 6 The system **BS4** consists of the axioms (A4)–(A8) added to **BS**.

Although we focus on the Hilbert-style presentation in this section, we may present **BS4** in terms of natural deduction as follows. This should be useful for readers who are more familiar with natural deduction.

Definition 7 $\mathscr{N}_1$**BS4** is the expansion of $\mathscr{N}$**BD** enriched by ($\circ_1$), ($\circ_2$), ($\circ_3$), ($\circ_4$), ($\circ \supset_1$), ($\circ \supset_2$), ($\sim \circ \supset$), ($\supset_1$) and ($\supset_2$), together with the introduction and elimination rules of $\supset$.

Proposition 1 *Given any set $\Sigma \cup \{A\}$ of formulas, $\Sigma \vdash_{\mathscr{N}_1\mathbf{BS4}} A$ iff $\Sigma \vdash_{\mathbf{BS4}} A$.*

The semantics we consider is given by the notion of four-valued valuation as in Definition 1. Based on this, we prove the following theorem.

Theorem 2 *A is four-valued tautology iff $\vdash_{\mathbf{BS4}} A$, for all formulas A of $\mathscr{L}_{\circ,\supset}$.*

The soundness part, i.e. to prove that all the theses of **BS4** are four-valued tautology, is easy as usual. Indeed, we only have to check that all the axioms take the designated values for any four-valued valuation for $\mathsf{Form}_{\circ,\supset}$, and that modus ponens preserves the designated values. Therefore, we will focus on the harder direction, completeness.

The method we employ here is the so-called Kalmár's method which can be found, for example, in [11, p. 42] for the case applied to classical propositional calculus. There are also some examples applied to systems of paraconsistent logic, such as [17] and [4]. In the former, the system known as $\mathbf{P}^1$ is proved to be complete with respect to a three-valued semantics, and in the latter two systems **LFI1** and **LFI2** are proved to be complete with respect to another three-valued semantics.

To begin with, we list some theses that are provable in **BS4**. Note here that $\neg$ defined in **BS** is already classical and so, any theses of classical propositional calculus containing classical negation hold in **BS4** as well.

(1) $((A \wedge \neg\sim A) \vee (\neg A \wedge \sim A)) \supset \circ A$ (2) $\circ A \supset \neg\sim\circ A$
(3) $((A \wedge \sim A) \vee (\neg A \wedge \neg\sim A)) \supset \neg\circ A$ (4) $\neg\circ A \supset \sim\circ A$
(5) $\neg\sim(A \supset B) \equiv (\neg A \vee \neg\sim B)$ (6) $\neg A \supset (A \supset B)$
(7) $A \supset (B \supset A)$ (8) $\neg(A \supset B) \equiv (A \wedge \neg B)$
(9) $(A \wedge \neg\sim A) \vee (A \wedge \sim A) \vee (\neg A \wedge \neg\sim A) \vee (\neg A \wedge \sim A)$

Lemma 1 *The following formulas are provable in the system* **BS4**.

Proof For (1), it suffices to prove $(A \wedge \neg\sim A) \supset \circ A$ and $(\neg A \wedge \sim A) \supset \circ A$. But these can be proved easily in view of equivalences $\circ A \equiv (A \equiv \neg\sim A)$ and $\circ A \equiv (\neg A \equiv \sim A)$ respectively which are both equivalent to the characteristic thesis $\circ A \equiv (\neg(A \wedge \sim A) \wedge (A \vee \sim A))$. As for (3), it suffices to prove $(A \wedge \sim A) \supset \neg\circ A$ and $(\neg A \wedge \neg\sim A) \supset \neg\circ A$. But these can be proved easily in view of the classical negation, defined in **BS4**, of the characteristic thesis. For (2) and (4), note first that the characteristic thesis is equivalent to $\circ A \equiv \neg(A \equiv \sim A)$, and therefore we have $\neg\circ A \equiv (A \equiv \sim A)$. In view of this equivalence, (A8) of **BS4** is equivalent to $\sim\circ A \equiv \neg\circ A$ which gives the desired results. (5) is easily proved by (A6) and the properties of classical negation defined in **BS4**, and (6) to (8) are well-known theses of classical propositional calculus. Finally, (9) is equivalent to the law of excluded middle with respect to classical negation $\neg$ defined in **BS4**. Indeed, the disjunction of the first two disjuncts are equivalent to $A \wedge (\neg\sim A \vee \sim A)$, hence equivalent to A since $\neg\sim A \vee \sim A$ is a special case of the law of excluded middle with respect to $\neg$. Likewise, the disjunction of the latter two disjuncts are equivalent to $\neg A$. Thus (9) is equivalent to $A \vee \neg A$, and therefore a thesis of **BS4**. □

Remark 2 Note that (1) and (3) is already provable in **BS** whereas the provability of other formulas related to $\sim$ depends on axioms unique to **BS4**.

By making use of this lemma, we obtain the following lemma which is the key for the completeness proof.

Lemma 2 *Given a four-valued valuation v, we define for each formula A an associated formula A^v:*

$$A^v = \begin{cases} A \wedge \neg\sim A & v(A) = \mathbf{t} \\ A \wedge \sim A & if\ \ v(A) = \mathbf{b} \\ \neg A \wedge \neg\sim A & v(A) = \mathbf{n} \\ \neg A \wedge \sim A & v(A) = \mathbf{f} \end{cases}$$

Now, let F be a formula whose set of atomic variables is $\{p_1, p_2, \ldots, p_n\}$, and let Δ^v be the set $\{p_1^v, p_2^v, \ldots, p_n^v\}$. Then, $\Delta^v \vdash F^v$.

Proof We proceed by induction on the number of connectives in F.
Base If $n = 0$, then F is p_i, so we need to show that $p_i^v \vdash p_i^v$, but this holds in **BS4**.
Induction Step Suppose that the desired result holds for the cases where the number of connective is less than $n + 1$. We show that it also holds in case of $n + 1$. We split

the cases depending on the main connective. Here, we only deal with cases where the main connectives are $\circ$ and $\supset$.

Case 1 If $F = \circ G$, then by induction hypothesis, we have $\Delta^v \vdash G^v$. We split the cases further depending on the value of G.

- If $v(G) = \mathbf{t}$ or $\mathbf{f}$, then we have $\Delta^v \vdash G \wedge \neg \sim G$ or $\Delta^v \vdash \neg G \wedge \sim G$ respectively, and hence $\Delta^v \vdash \circ G \wedge \neg \sim \circ G$ by (1) and (2). On the other hand, by the definition of the valuation, we have $v(\circ G) = \mathbf{t}$, namely $v(F) = \mathbf{t}$. This shows that the desired result holds. Indeed, $\Delta^v \vdash \circ G \wedge \neg \sim \circ G$ is $\Delta^v \vdash F \wedge \sim \neg F$, and therefore $\Delta^v \vdash F^v$.
- If $v(G) = \mathbf{b}$ or $\mathbf{n}$, then we have $\Delta^v \vdash G \wedge \sim G$ or $\Delta^v \vdash \neg G \wedge \neg \sim G$ respectively, and hence $\Delta^v \vdash \neg \circ G \wedge \sim \circ G$ by (3) and (4). On the other hand, by the definition of the valuation, we have $v(\circ G) = \mathbf{f}$, namely $v(F) = \mathbf{f}$. This shows that the desired result holds. Indeed, $\Delta^v \vdash \neg \circ G \wedge \sim \circ G$ is $\Delta^v \vdash \neg F \wedge \sim F$, and therefore $\Delta^v \vdash F^v$.

Case 2 If $F = G \supset H$, then by induction hypothesis, we have $\Delta^v \vdash G^v$ and $\Delta^v \vdash H^v$. We split the cases further depending on the values of G and H.

- If $v(G) = \mathbf{f}$ or $v(G) = \mathbf{n}$ or $v(H) = \mathbf{t}$, then we have $\Delta^v \vdash \neg G \wedge \sim G$ or $\Delta^v \vdash \neg G \wedge \neg \sim G$ or $\Delta^v \vdash H \wedge \neg \sim H$ respectively. Since the first two cases imply $\Delta^v \vdash \neg G$, we obtain $\Delta^v \vdash \neg G$ or $\Delta^v \vdash H \wedge \neg \sim H$. Then, by (6) or (7) respectively together with (5), we obtain $\Delta^v \vdash (G \supset H) \wedge \neg \sim (G \supset H)$. On the other hand, by the definition of the valuation, we have $v(G \supset H) = \mathbf{t}$, namely $v(F) = \mathbf{t}$. This shows that the desired result holds. Indeed, $\Delta^v \vdash (G \supset H) \wedge \neg \sim (G \supset H)$ is $\Delta^v \vdash F \wedge \neg \sim F$, and therefore $\Delta^v \vdash F^v$.
- If $v(G) = \mathbf{t}$ or $\mathbf{b}$, then we have $\Delta^v \vdash G \wedge \neg \sim G$ or $\Delta^v \vdash G \wedge \sim G$ respectively, but in both cases, we obtain $\Delta^v \vdash G$.
  - If $v(H) = \mathbf{b}$, then we have $\Delta^v \vdash H \wedge \sim H$ and therefore $\Delta^v \vdash H \wedge (G \wedge \sim H)$ which implies $\Delta^v \vdash (G \supset H) \wedge \sim (G \supset H)$ by (7) and (A6).
  - If $v(H) = \mathbf{n}$, then we have $\Delta^v \vdash \neg H \wedge \neg \sim H$ and therefore $\Delta^v \vdash (G \wedge \neg H) \wedge \neg \sim H$ which implies $\Delta^v \vdash \neg (G \supset H) \wedge \neg \sim (G \supset H)$ by (8) and (5).
  - If $v(H) = \mathbf{f}$, then we have $\Delta^v \vdash \neg H \wedge \sim H$ and therefore $\Delta^v \vdash (G \wedge \neg H) \wedge (G \wedge \sim H)$ which implies $\Delta^v \vdash \neg (G \supset H) \wedge \sim (G \supset H)$ by (8) and (A6).

  On the other hand, in each case, we have $v(G \supset H) = \mathbf{b}$ or $\mathbf{n}$ or $\mathbf{f}$, namely $v(F) = \mathbf{b}$ or $\mathbf{n}$ or $\mathbf{f}$ by the definition of the valuation respectively. These show that the desired results hold. Indeed, for the case $v(H) = \mathbf{b}$, $\Delta^v \vdash (G \supset H) \wedge \sim (G \supset H)$ is $\Delta^v \vdash F \wedge \sim F$, and therefore $\Delta^v \vdash F^v$.

This completes the proof. □

By making use of this lemma, we prove the completeness part of Theorem 2.

Proof of Theorem 2 Let F be any four-valued tautologies and Δ be the set of propositional variables occurring in F. Then by the previous lemma, we have $\Delta^v \vdash F^v$. Furthermore, since F is a four-valued tautology, one of $\Delta^v \vdash (F \wedge \neg\sim F)$ or $\Delta^v \vdash (F \wedge \sim F)$ holds. Therefore, in both cases, we obtain $\Delta^v \vdash F$.

Now, let Δ^v_k be the set $\Delta^v \setminus p_k$, and suppose that four valuations v_1, v_2, v_3 and v_4 be those that satisfy $\Delta^{v_1}_k = \Delta^{v_2}_k = \Delta^{v_3}_k = \Delta^{v_4}_k (=_{\text{def.}} \Delta_k)$ and $v_1(p_k) = \mathbf{t}$, $v_2(p_k) = \mathbf{b}$, $v_3(p_k) = \mathbf{n}$ and $v_4(p_k) = \mathbf{f}$. Then, for v_1, $\Delta^v \vdash F$ is $\Delta^{v_1}_k, \{p_k \wedge \neg\sim p_k\} \vdash F$ by the definition of A^v, and therefore, by the deduction theorem, we have $\Delta^{v_1}_k \vdash (p_k \wedge \neg\sim p_k) \supset F$. Similarly, we obtain $\Delta^{v_2}_k \vdash (p_k \wedge \sim p_k) \supset F$, $\Delta^{v_3}_k \vdash (\neg p_k \wedge \neg\sim p_k) \supset F$ and $\Delta^{v_4}_k \vdash (\neg p_k \wedge \sim p_k) \supset F$ for v_2, v_3 and v_4 respectively. Putting these results together by the fact that $\Delta^{v_1}_k = \Delta^{v_2}_k = \Delta^{v_3}_k = \Delta^{v_4}_k = \Delta_k$, we have $\Delta_k \vdash ((p_k \wedge \neg\sim p_k) \vee (p_k \wedge \sim p_k) \vee (\neg p_k \wedge \neg\sim p_k) \vee (\neg p_k \wedge \sim p_k)) \supset F$. Since we have (9), we may conclude that $\Delta_k \vdash F$. And after repeating this procedure for $k - 1$ more times, we obtain $\vdash F$ which is the desired result. □

Before turning to further considerations on the classicality operator of **BS4**, we will briefly consider a consistency operator in four-valued logic. The truth table for the classicality operator $\circ$ reflects our understanding of da Costa's idea well, since it clearly distinguishes 'classical' values **t** and **f** from 'non-classical' values **b** and **n**. And in a similar manner, it is also possible to consider the notion of a consistency operator as well. There are several possibilities, but a simple way to distinguish the consistency and the inconsistency is to consider the following operation:

A	$\circ' A$
t	**t**
b	**f**
n	**t**
f	**t**

This is because the only inconsistent value is **b**, and the above operator $\circ'$ clearly distinguishes the inconsistent value from other consistent values.

Then, what is the relation between the classicality operator $\circ$ and the consistency operator $\circ'$? One of the answers is that their expressive power is equivalent if we assume the truth tables for $\sim$, $\wedge$, $\vee$ and $\supset$. This may be easily observed by the fact that both truth tables are equivalent to the truth table enriched by $\perp$ that satisfies the following:

A	$\perp$
t	**f**
b	**f**
n	**f**
f	**f**

Indeed, on the one hand, $\circ A \wedge A \wedge \sim A$ and $\circ' A \wedge A \wedge \sim A$ define $\bot$ in both expansions by $\circ$ and $\circ'$ respectively. On the other hand, if we have $\bot$, then we can define the strong negation $\neg A$ of A by $A \supset \bot$. By making use of this strong negation, $\neg(A \wedge \sim A) \wedge (A \vee \sim A)$ and $\neg(A \wedge \sim A)$ define $\circ$ and $\circ'$ respectively.

Another point to be noted is the relation between $\circ'$ and the formula $\sim(A \wedge \sim A)$ which was the key formula in da Costa's system $\mathbf{C}_1$. In some of the truth tables such as one of the four-valued generalizations of the truth tables[7] in [6, p. 499], $\sim(A \wedge \sim A)$ indeed defines $\circ'$. However, this is not the case in our truth tables, as one can easily observe. The crucial point is that when the formula $\sim(A \wedge \sim A)$ is given a special role to control the behavior of contradictions just as the consistency of A in LFIs,[8] it is equivalent to the formula $\neg(A \wedge \sim A)$ where $\neg$ is strong negation.[9] And if we consider the latter formula, then this will define $\circ'$ in our truth tables, as we have observed above.

In view of these facts, we may develop the proof theory of systems with $\circ'$ as well, but we shall keep the detail for another paper. And instead, we will continue our investigation on the classicality operator of **BS4**.

4 Do We Really Need the Material Conditional?

Now, having the system **BS4** above, one of the natural questions from those following Belnap-Dunn tradition must be as follows: is the material conditional essential in formalizing the idea of classicality operator? Recall here that the notion of classicality helps us in clarifying the area where we can apply inferences of classical logic. Since some of those including Priest and Beall do not hesitate to make use of classical inferences when it is safe enough to do so, the idea of da Costa must be of interest. All this being said, the answer to the above question is in the affirmative. In other words, we can enrich the system of Belnap-Dunn with classicality operator *without* being forced to accept the material conditional. Our plan for this section is as follows. First, by making use of the natural deduction formulation of **BS4**, we introduce a system without the material conditional in the systems. Second, we prove the completeness of this system with respect to a semantics obtained by adding a classicality operator to Belnap-Dunn's four-valued semantics. And finally, we show that material conditional cannot be defined in the semantics.

To begin with, we consider an alternative formulation of **BS4** in terms of natural deduction. While the rules $(\circ \supset_1)$, $(\circ \supset_2)$ and $(\sim \circ \supset)$ of $\mathscr{N}_1$**BS4** contain both the

---

[7] The truth tables there are known as the truth tables for the system $\mathbf{P}^1$ of Sette studied in [17].

[8] Namely, $\sim(A \wedge \sim A) \supset (A \supset (\sim A \supset B))$ holds, just like (A2) of **cBS**.

[9] The outline of the proof for the equivalence is as follows. First, $\neg(A \wedge \sim A) \supset \sim(A \wedge \sim A)$ follows as a special case of $\neg A \supset \sim A$, and this is equivalent to $A \vee \sim A$ which is assumed in concerned systems. For the other way around, we assume $\sim(A \wedge \sim A)$ and $(A \wedge \sim A)$. Then, by the special role given to $\sim(A \wedge \sim A)$, the conjunction of the formulas are explosive. So, we have $\sim(A \wedge \sim A) \wedge (A \wedge \sim A) \supset \neg(A \wedge \sim A)$ in particular, and finally by reductio with respect to $\neg$, we obtain the desired formula. For the stronger definition, see Theorem 1.

connectives $\circ$ and $\supset$, we can reformulate the rule set for $\circ$ and $\supset$ of $\mathcal{N}_1$**BS4** into the rule set of $\mathcal{N}_2$**BS4** where the behaviors of those connectives are modular.

Definition 8 $\mathcal{N}_2$**BS4** consists of ($\circ_1$), ($\circ_2$), ($\circ_3$), ($\circ_4$), ($\circ_5$), ($\supset_1$) and ($\supset_2$) from Table 2 plus the rules of $\mathcal{N}$**BD** and the introduction and elimination rules of $\supset$.[10]

Proposition 2 *Given any set* $\Sigma \cup \{A\}$ *of formulas,* $\Sigma \vdash_{\mathcal{N}_1\mathbf{BS4}} A$ *iff* $\Sigma \vdash_{\mathcal{N}_2\mathbf{BS4}} A$.

Proof First, note that it follows from ($\circ_5$) that $\circ A, {\sim}\circ A \vdash_{\mathcal{N}_2\mathbf{BS4}} B$ (by ($\circ_1$)) and $\vdash_{\mathcal{N}_2\mathbf{BS4}} \circ A \vee {\sim}\circ A$ (by ($\circ_2$)). Then, the left-to-right direction is easy to establish, and so, let us concentrate on the right-to-left direction here. The hardest part consists in showing the derivability of ($\circ_5$) $\circ\circ A$ in $\mathcal{N}_1$**BS4**. In view of the characteristic thesis (1) of $\circ$, it suffices to demonstrate that $\circ A, {\sim}\circ A \vdash_{\mathcal{N}_1\mathbf{BS4}} B$ and $\vdash_{\mathcal{N}_1\mathbf{BS4}} \circ A \vee {\sim}\circ A$. The former is established as follows.

$$\dfrac{\dfrac{\circ A}{A \vee {\sim} A}(\circ_2) \qquad \dfrac{\dfrac{{\sim}\circ A \quad [A]_1}{{\sim} A}(\circ_3) \quad [A]_1 \quad \circ A}{B}(\circ_1) \qquad \dfrac{\dfrac{{\sim}\circ A \quad [{\sim} A]_2}{A}(\circ_4) \quad [{\sim} A]_2 \quad \circ A}{B}(\circ_1)}{B}1,2$$

Let us move to the latter. Since $\neg$ is the classical negation defined in **BS4**, note that we can use (RAA). Then, we proceed as follows. Assume $\neg(\circ A \vee {\sim}\circ A)$. Then, we can derive $\neg\circ A$ and $\neg{\sim}\circ A$, since $\neg$ is the classical negation defined in **BS4**. By Lemma 1 (6) and $\neg\circ A$, we obtain ${\sim}\circ A$. Together with $\neg{\sim}\circ A$, we get $\bot$. (RAA) tells us that $(\circ A \vee {\sim}\circ A)$ by discharging the initial assumption $\neg(\circ A \vee {\sim}\circ A)$. This finishes the proof of the right-to-left direction. □

This proposition allows us to obtain the natural deduction calculi which is the conditional $\supset$-free fragment of the underlying syntax of **BS4**.

Definition 9 $\mathcal{N}$**BD**$\circ$ consists of ($\circ_1$), ($\circ_2$), ($\circ_3$), ($\circ_4$), ($\circ_5$) from Table 2 in addition to all the rules of $\mathcal{N}$**BD**.

As for the semantics for $\mathcal{L}_\circ$, we make use of the notions of four-valued valuation and semantic consequence from Definitions 1 and 2. Now we turn to the completeness of $\mathcal{N}$**BD**$\circ$.

Theorem 3 *Given any set* $\Sigma \cup \{A\}$ *of formulas,* $\Sigma \vdash_{\mathcal{N}\mathbf{BD}\circ} A$ *iff* $\Sigma \models_4 A$.

Proof (Outline) For convenience, we reformulate our semantics based on truth tables for $\mathcal{L}_\circ$ in terms of the *positive* and *negative* clauses as follows. We interpret **t**, **b**, **n**, **f** as $\{1\}$, $\{0, 1\}$, $\emptyset$, $\{0\}$, respectively. Then, we can rewrite the truth table of $\circ A$ by the following two clauses:

$$1 \in v(\circ A) \text{ iff } (1 \in v(A) \text{ and } 0 \notin v(A)) \text{ or } (1 \notin v(A) \text{ and } 0 \in v(A))$$
$$0 \in v(\circ A) \text{ iff } (1 \in v(A) \text{ iff } 0 \in v(A))$$

[10] More results on related systems of $\mathcal{N}_2$**BS4** is proved in [16].

Moreover, we can show that $\Sigma \models_4 B$ iff, for all valuations v such that $1 \in v(C)$ for all $C \in \Sigma$, $1 \in v(B)$. Then, let us establish direction corresponding to the completeness, i.e., the right-to-left direction. We show the contrapositive implication. So, suppose that $\Sigma \nvdash_{\mathcal{N}\mathbf{BD}\circ} A$. Then, we can find a *prime theory* Σ^+ (with respect to $\mathcal{N}\mathbf{BD}\circ$) such that $\Sigma \subseteq \Sigma^+$ and $A \notin \Sigma^+$. Let us define $v_{\Sigma^+} : \mathsf{Var} \to \mathscr{P}(\{0, 1\})$ as follows:

$$1 \in v_{\Sigma^+}(p) \text{ iff } p \in \Sigma^+, \qquad 0 \in v_{\Sigma^+}(p) \text{ iff } {\sim} p \in \Sigma^+.$$

Now, what remains to be proved is the equivalences:

$$1 \in v_{\Sigma^+}(C) \text{ iff } C \in \Sigma^+, \qquad 0 \in v_{\Sigma^+}(C) \text{ iff } {\sim} C \in \Sigma^+.$$

Indeed, with this result at hand, we may conclude that $1 \in v_{\Sigma^+}(C)$ for all $C \in \Sigma$ but $1 \notin v_{\Sigma^+}(A)$, i.e., $\Sigma \not\models_4 A$ as desired. Here we only check the above equivalences for the case where C is of the form $\circ B$. For the positive clause, we proceed as follows.

$$\begin{aligned} 1 \in v_{\Sigma^+}(\circ B) &\text{ iff } (1 \in v_{\Sigma^+}(B) \text{ and } 0 \notin v_{\Sigma^+}(B)) \text{ or } (0 \in v_{\Sigma^+}(B) \text{ and } 1 \notin v_{\Sigma^+}(B)) \\ &\text{ iff } (B \in \Sigma^+ \text{ and } {\sim} B \notin \Sigma^+) \text{ or } ({\sim} B \in \Sigma^+ \text{ and } B \notin \Sigma^+) \quad \text{(by I.H.)} \end{aligned}$$

We need to show that this last line is equivalent with $\circ B \in \Sigma^+$. Assume that the last line holds. Consider the disjunct $B \in \Sigma^+$ and ${\sim} B \notin \Sigma^+$. Moreover, suppose for contradiction that $\circ B \notin \Sigma^+$. Since $\vdash_{\mathcal{N}\mathbf{BD}\circ} \circ B \vee {\sim} \circ B$, we have ${\sim} \circ B \in \Sigma^+$. Then, the rule ($\circ_3$) and $B \in \Sigma^+$ jointly imply ${\sim} B \in \Sigma^+$, which contradicts ${\sim} B \notin \Sigma^+$. We can offer the similar argument for the other disjunct. Conversely, suppose that $\circ B \in \Sigma^+$. We need to demonstrate the last line of the displayed statements. Suppose for contradiction that the last line fails, i.e., ($B \in \Sigma^+$ implies ${\sim} B \in \Sigma^+$) and (${\sim} B \in \Sigma^+$ implies $B \in \Sigma^+$). By the rule ($\circ_2$) and our assumption $\circ B \in \Sigma^+$, we obtain $B \vee {\sim} B \in \Sigma^+$, which implies $B \in \Sigma^+$ or ${\sim} B \in \Sigma^+$. Here we concentrate on the disjunct ${\sim} B \in \Sigma^+$. By our counterfactual assumption, we obtain $B \in \Sigma^+$. By the rule ($\circ_1$) and $B, {\sim} B, \circ B \in \Sigma^+$, we get $A \in \Sigma^+$, which contradicts $A \notin \Sigma^+$.

Let us move to the negative clause. First of all, note that $0 \in v_{\Sigma^+}(\circ B)$ iff $1 \notin v_{\Sigma^+}(\circ B)$. Then, it suffices to show that $\circ B \notin \Sigma^+$ iff ${\sim} \circ B \in \Sigma^+$, in order to establish the negative clause. However, this is an easy consequence of the following in $\mathcal{N}\mathbf{BD}\circ$: $\vdash_{\mathcal{N}\mathbf{BD}\circ} \circ B \vee {\sim} \circ B$ and $\circ B, {\sim} \circ B \vdash_{\mathcal{N}\mathbf{BD}\circ} A$. □

Finally, we prove that the material conditional is not definable in $\mathcal{N}\mathbf{BD}\circ$ by making use of the semantics. As a first step, we make explicit what we understand by classical negation and the material conditional with the help of the notion of four-valued semantic consequence from Definition 2.

Definition 10 Let $\models_4$ be a semantic consequence relation from Definition 2. Then,

(i) Any unary connective $\neg$ which satisfies $A \wedge \neg A \models_4 B$ and $B \models_4 A \vee \neg A$ is referred to as a *classical negation*,

(ii) Any binary connective $\rightarrow$ which satisfies $A \wedge (A \rightarrow B) \models_4 B$ and $C \models_4 A \vee (A \rightarrow B)$ is referred to as a *material conditional*.

Remark 3 The intuitions behind the above definitions are as follows. First, in one of the formulations of classical propositional calculus, *ex contradictione quodlibet* and the law of excluded middle are the formulas characterizing classical negation, and this fact is reflected in (i). Second, there are some formulations of classical propositional calculus not having negation as a primitive connective, but defined by the "arrow-bottom method" (cf. Lemma 3). And from such a viewpoint, conditions in (ii) correspond to the conditions in (i). It is also worth noting that $A \vee (A \rightarrow B)$ occurring in (ii) is the formula that fills the gap between **IPC**$^+$ and **CPC**$^+$ (cf. Definition 4).

Remark 4 If we reformulate the above conditions for classical negation, then we obtain the equivalence $v(\neg A) \in \mathcal{D}$ iff $v(A) \notin \mathcal{D}$. As for the conditions for the material conditional, we do *not* have the equivalence $v(A \supset B) \in \mathcal{D}$ iff $v(A) \notin \mathcal{D}$ or $v(B) \in \mathcal{D}$, since we do not have the implication if $v(B) \in \mathcal{D}$ then $v(A \supset B) \in \mathcal{D}$.

Remark 5 Note that, as we can easily verify, the negation $\neg$ and $\supset$ in **BS4** both satisfy the above conditions respectively. But these two are only examples and there are more connectives which satisfy the above conditions. In other words, the above conditions do not necessarily determine the unary or binary connective uniquely.

Our purpose is to show the undefinability of material conditional in $\mathcal{L}_\circ$, but it is actually sufficient to consider the definability of classical negation, according to the following lemma:

Lemma 3 *If a material conditional is definable in $\mathcal{L}_\circ$, so is a classical negation.*

Proof Note that we have the bottom particle $\bot$ defined as $\bot =_{\text{def.}} \circ X \wedge X \wedge {\sim} X$ for some X in $\mathcal{L}_\circ$. Therefore, the negation $\neg$ defined as $\neg A =_{\text{def.}} A \supset \bot$ will satisfy the above conditions for the classical negations. □

Remark 6 First, note that the above lemma does not hold in $\mathcal{L}$, i.e., the syntax with $\sim$, $\wedge$, and $\vee$ alone, since the bottom particle is not definable in $\mathcal{L}$. Second, one may expect that, by following the convention in da Costa's systems, ${\sim} A \wedge \circ A$ will satisfy the conditions for classical negations. However, this is not the case since the second condition $A \models_4 B \vee \neg B$ is not met when we assign the values **n** and **t** to A and B respectively.

The key for our desired result is the following lemma.

Lemma 4 *Let $A(p)$ be a formula of $\mathcal{L}_\circ$ which contains p as the only propositional variable in A. Then one of the following holds:*

(i) $v(A(p)) = \mathbf{f}$ *when* $v(p) \in \{\mathbf{b}, \mathbf{n}\}$ *for any four-valued valuation* v.
(ii) $v(A(p)) = \mathbf{t}$ *when* $v(p) \in \{\mathbf{b}, \mathbf{n}\}$ *for any four-valued valuation* v.
(iii) $v(A(p)) = v(p)$ *when* $v(p) \in \{\mathbf{b}, \mathbf{n}\}$ *for any four-valued valuation* v.

Proof We proceed by induction on the complexity of $A(p)$.
Base If $A(p)$ is p, then it satisfies the condition (iii).
Induction step We split into four cases depending on the main connective.

- If $A(p) = \sim B(p)$ or $\circ\, B(p)$, then $B(p)$ satisfies one of the three conditions by induction hypothesis. And with the truth table for $\sim$ and $\circ$ in mind, $A(p)$ behaves as in the table below.
- If $A(p) = B(p) \wedge C(p)$ or $B(p) \vee C(p)$, then $B(p)$ and $C(p)$ both satisfy one of the three conditions by induction hypothesis. And with the truth table for $\wedge$ and $\vee$ in mind, $A(p)$ behaves as in the table below.

$B(p)$	$\sim B(p)$	$\circ B(p)$
(i)	(ii)	(ii)
(ii)	(i)	(ii)
(iii)	(iii)	(i)

$B(p) \wedge C(p)$	(i)	(ii)	(iii)
(i)	(i)	(i)	(i)
(ii)	(i)	(ii)	(iii)
(iii)	(i)	(iii)	(iii)

$B(p) \vee C(p)$	(i)	(ii)	(iii)
(i)	(i)	(ii)	(iii)
(ii)	(ii)	(ii)	(ii)
(iii)	(iii)	(ii)	(iii)

This completes the proof. □

Theorem 4 *Neither classical negation nor the material conditional is definable in $\mathscr{L}_\circ$. In particular, $\neg$ and $\supset$ of Table 1 are not definable in $\mathscr{L}_\circ$.*

Proof In view of Lemma 3, it suffices to show that any of the classical negations cannot be defined in $\mathscr{L}_\circ$. So, let us suppose that a classical negation $\neg$ is definable. Then, in view of Lemma 4, $\neg$ will satisfy one of the three conditions (i)–(iii).

- If (i) is satisfied then we have in particular that if $v(p) = \mathbf{n}$ then $v(\neg p) = \mathbf{f}$, namely if $v(p) \notin \mathscr{D}$ then $v(\neg p) \notin \mathscr{D}$, which is impossible in view of Remark 4.
- If (ii) is satisfied then we have in particular that if $v(p) = \mathbf{b}$ then $v(\neg p) = \mathbf{t}$, namely if $v(p) \in \mathscr{D}$ then $v(\neg p) \in \mathscr{D}$, which is impossible in view of Remark 4.
- If (iii) is satisfied then we have $v(\neg p) = v(p)$, which is impossible in view of Remark 4.

Thus, classical negation is not definable in $\mathscr{L}_\circ$. □

Remark 7 Note that if we assume LEM with respect to $\sim$ in $\mathscr{N}$BDo, then we can define classical negation as $\sim A \wedge \circ A$. In other words, if we add classicality (or consistency) to **LP** of Priest (cf. [14]), then we obtain the system equivalent to **LFI1** of Carnielli, Marcos and de Amo.

Although our non-definability proof was given in a purely semantic manner, we may also prove the result in terms of proof theory together with the completeness result. For this purpose, we need a proof-theoretical version of Lemma 4.

Lemma 5 *Let $A(p)$ be a formula of $\mathscr{L}_\circ$ which contains p as the only propositional variable in A. Then one of the following holds:*

(i) $\sim\circ p \vdash_{\mathcal{N}\mathbf{BD}\circ} \sim A(p)$ *and* $\sim\circ p, A(p) \vdash_{\mathcal{N}\mathbf{BD}\circ} \bot$.
(ii) $\sim\circ p \vdash_{\mathcal{N}\mathbf{BD}\circ} A(p)$ *and* $\sim\circ p, \sim A(p) \vdash_{\mathcal{N}\mathbf{BD}\circ} \bot$.
(iii) $\sim\circ p, p \vdash_{\mathcal{N}\mathbf{BD}\circ} A(p)$, $\sim\circ p, A(p) \vdash_{\mathcal{N}\mathbf{BD}\circ} p$, $\sim\circ p, \sim p \vdash_{\mathcal{N}\mathbf{BD}\circ} \sim A(p)$ *and* $\sim\circ p, \sim A(p) \vdash_{\mathcal{N}\mathbf{BD}\circ} \sim p$.

Proof By rewriting Lemma 4 by the completeness result (Theorem 3), we obtain the desired statement. Or, we may directly proceed by induction on the complexity of $A(p)$. □

And by making use of this lemma and the completeness result, we may prove the non-definability as follows.

Alternative proof of Theorem 4 In view of Lemma 3, it suffices to show that any of the classical negations cannot be defined in $\mathcal{L}_\circ$. So, let us suppose that a classical negation $\neg$ is definable. Then, in view of Lemma 4, $\neg$ will satisfy one of the three conditions (i)–(iii).

- If (i) is satisfied then we have in particular that $\sim\circ p, \neg p \vdash_{\mathcal{N}\mathbf{BD}\circ} \bot$. On the other hand, consider a four-valued valuation v_0 such that $v_0(p) = \mathbf{n}$. Then, we have $v_0(\sim\circ p) = \mathbf{t} \in \mathcal{D}$ by the truth table and $v_0(\neg p) \in \mathcal{D}$ by Remark 4. We also have $v_0(\bot) = \mathbf{f} \notin \mathcal{D}$, so we obtain $\sim\circ p, \neg p \not\models_4 \bot$, and therefore $\sim\circ p, \neg p \nvdash_{\mathcal{N}\mathbf{BD}\circ} \bot$ by soundness. But this is a contradiction.
- If (ii) is satisfied then the proof is analogous to the previous case. In particular, we find a contradiction with $\sim\circ p \vdash_{\mathcal{N}\mathbf{BD}\circ} \neg p$.
- If (iii) is satisfied then we have in particular that $\sim\circ p, p \vdash_{\mathcal{N}\mathbf{BD}\circ} \neg p$ and $\sim\circ p, \neg p \vdash_{\mathcal{N}\mathbf{BD}\circ} p$. Since $\neg$ is a classical negation, the completeness result enables us to obtain $\sim\circ p \vdash_{\mathcal{N}\mathbf{BD}\circ} \bot$. However, by a similar argument using soundness, we can establish $\sim\circ p \nvdash_{\mathcal{N}\mathbf{BD}\circ} \bot$ as well, which is a contradiction.

This completes the proof. □

5 Can We Place Classicality Operator in Constructive Context?

In view of the results obtained in the previous section, the classicality operator of **BS4** is not necessarily accompanied by any of the material conditionals we defined (cf. Definition 10). But then, we may question if we can generalize the classicality operator where the conditional is taken to be intuitionistic (or constructive). We answer this question affirmatively by developing a system that has a close relation to a system of constructive negation, whose origin can be traced back to the work of Nelson (cf. [9, 12]). Note, however, that our attempt is not the very first in trying to place the notion of classicality operator in a constructive setting. Indeed, there is an attempt by Guillaume (cf. [7]) which considers a constructive counterpart of the basic systems of LFIs. Still, his work does not touch the relation to the systems of Nelson, so in this regard, the present paper will be the first to reveal the relation between the ideas of da Costa and Nelson. We will start by revisiting the system of Nelson.

Definition 11 **N4** consists of axioms (A4)–(A7) together with $\mathbf{IPC}^+$ (cf. Definition 4.), and $\mathbf{N4}^\bot$ is obtained by adding two axioms $\bot \supset A$ and $A \supset {\sim}\bot$ to **N4**.

The original idea was given by Nelson back in the forties, and various systems within the idea of Nelson has been studied. Many of them were paracomplete but not paraconsistent, though two systems, **N4** and its extension $\mathbf{N4}^\bot$, are both paracomplete and paraconsistent.

Now, in [13], a translation theorem between the system **BS4** and an extension of the system $\mathbf{N4}^\bot$ enriched by Peirce's law, referred to as $\mathbf{B4}^\bot$, was established. However, the problem of finding the relation between a constructive version of **BS4** and $\mathbf{N4}^\bot$ was left open. Note here that it must be of interest for those in the tradition of Nelson to see if we can introduce the notion of classicality operator, which makes clear the area we can apply the classical inference. Therefore the above question must be quite natural and worth exploring. Based on these, the purpose of this section is to show that we can prove a translation theorem between a constructive version of **BS4** and $\mathbf{N4}^\bot$. For this purpose, we now introduce the constructive version of **BS4**.

Definition 12 The system **cBS4** consists of the axioms (A4)–(A8) added to **cBS** (cf. Definition 4.). Equivalently, **cBS4** is obtained by adding axioms (A1)–(A3) and (A8) to **N4**.

The translations to be made use of in the following are as follows:

Definition 13 Let τ_1 be a translation from **cBS4** to $\mathbf{N4}^\bot$ which satisfies the following conditions:

$$\begin{array}{ll}
\tau_1(p_i) = p_i & \tau_1({\sim} p_i) = {\sim} p_i \\
\tau_1(A \supset B) = \tau_1(A) \supset \tau_1(B) & \tau_1({\sim}(A \supset B)) = \tau_1(A) \wedge \tau_1({\sim} B) \\
\tau_1(A \wedge B) = \tau_1(A) \wedge \tau_1(B) & \tau_1({\sim}(A \wedge B)) = \tau_1({\sim} A) \vee \tau_1({\sim} B) \\
\tau_1(A \vee B) = \tau_1(A) \vee \tau_1(B) & \tau_1({\sim}(A \vee B)) = \tau_1({\sim} A) \wedge \tau_1({\sim} B) \\
\tau_1({\sim}{\sim} A) = \tau_1(A) & \tau_1({\sim}(\circ A)) = \tau_1(A) \equiv \tau_1({\sim} A) \\
\multicolumn{2}{l}{\tau_1(\circ A) = ((\tau_1(A) \wedge \tau_1({\sim} A)) \supset \bot) \wedge (\tau_1(A) \vee \tau_1({\sim} A))}
\end{array}$$

Also let τ_2 be a translation from $\mathbf{N4}^\bot$ to **cBS4** which satisfies the following conditions together with similar conditions for the translation of $A * B$ and ${\sim}(A * B)$ where $* \in \{\supset, \wedge, \vee\}$:

$$\begin{array}{lll}
\tau_2(p_i) = p_i & \tau_2({\sim} p_i) = {\sim} p_i & \tau_2({\sim}{\sim} A) = \tau_2(A) \\
\tau_2(\bot) = \circ p_1 \wedge p_1 \wedge {\sim} p_1 & \tau_2({\sim}\bot) = p_1 \supset p_1 &
\end{array}$$

Remark 8 Compared to the translations employed in [13], the only difference lies in the image of $\circ A$ under τ_1 which was set to be as: $\tau_1(\circ A) = (\tau_1(A) \equiv \tau_1({\sim} A)) \supset \bot$. The two formulas are obviously equivalent in $\mathbf{B4}^\bot$ but not in $\mathbf{N4}^\bot$, and this difference turned out to be the key in showing the desired result. Note that the translation here reflects the characteristic thesis (1) more directly compared to the previous version.

For the purpose of proving the desired result, the following lemma is the key.

Lemma 6 *Let τ_1 and τ_2 be translations defined above. Then,*

(i) If $\vdash_{\mathbf{cBS4}} A$ then $\vdash_{\mathbf{N4}^{\perp}} \tau_1(A)$ (iii) $\vdash_{\mathbf{cBS4}} \tau_2(\tau_1(A)) \equiv A$
(ii) If $\vdash_{\mathbf{N4}^{\perp}} B$ then $\vdash_{\mathbf{cBS4}} \tau_2(B)$ (iv) $\vdash_{\mathbf{N4}^{\perp}} \tau_1(\tau_2(B)) \equiv B$

Proof For (i) and (ii), it is sufficient to check that all the axioms of **cBS4** and $\mathbf{N4}^{\perp}$ translated by τ_1 and τ_2 are provable in $\mathbf{N4}^{\perp}$ and **cBS4** respectively, and that modus ponens remains valid under translation. Since **cBS4** and $\mathbf{N4}^{\perp}$ share most of the axioms, we only have to check few cases. Indeed, for the case of (i), we only have to verify that images of axioms (A1) to (A3) and (A8) by τ_1 are provable in $\mathbf{N4}^{\perp}$. These are straightforward; just note that both $\tau_1(A \supset \circ A)$ and $\tau_1({\sim} A \supset \circ A)$ are equivalent to $(\tau_1(A) \wedge \tau_1({\sim} A)) \supset \perp$. For the case of (ii), we need to verify that images of axioms $\perp \supset A$ and $A \supset {\sim} \perp$ by τ_2 are provable in **cBS4**, but these are obvious. As for (iii) and (iv), the proof is by induction on the complexity of A and B respectively. □

Theorem 5 *Let τ_1 and τ_2 be translations defined above. Then,*

$$\vdash_{\mathbf{cBS4}} A \text{ iff } \vdash_{\mathbf{N4}^{\perp}} \tau_1(A) \quad \text{and} \quad \vdash_{\mathbf{N4}^{\perp}} B \text{ iff } \vdash_{\mathbf{cBS4}} \tau_2(B).$$

Proof We only prove the former as the latter can be proved in a similar manner. Now, the right to the left direction is already proved in (i) of the previous lemma. For the other direction, suppose $\vdash_{\mathbf{N4}^{\perp}} \tau_1(A)$. Then by (ii) of the previous lemma, we obtain $\vdash_{\mathbf{cBS4}} \tau_2(\tau_1(A))$, but in view of (iii) of the previous lemma, we have $\vdash_{\mathbf{cBS4}} A$ which is the desired result. For the case of proving the latter, we need (ii), (i) and (iv) instead of (i), (ii) and (iii) respectively. □

We now turn to the semantics of **cBS4**. Let W be a non-empty set, $\leq$ a pre-order on W. Let V be a pair (V^+, V^-) of $V^+, V^- : \mathsf{Var} \to \mathscr{P}(W)$ that satisfy the *persistency*: $w \in V^*(p)$ and $w \leq w'$ imply $w' \in V^*(p)$ for all $w, w' \in W$ and all $p \in \mathsf{Var}$ (where $*$ is $+$ or $-$). Then, we say $(W, \leq, V)$ is a *Kripke model*. Given such a model, we define the satisfaction pair $(\models^+, \models^-)$ as usual except[11]:

$$w \models^+ \circ A \text{ iff } \forall w' \geq w\, ((w' \models^+ A \text{ and } w' \not\models^- A) \text{ or } (w \not\models^+ A \text{ and } w \models^- A));$$
$$w \models^- \circ A \text{ iff } \forall w' \geq w\, (w' \models^+ A \text{ iff } w' \models^- A).$$

Let us say that A is a *semantic* **c***-consequence* from Σ (notation : $\Sigma \models_{\mathbf{c}} A$) if, for all Kripke models $(W, \leq, V)$ and all $w \in W$ such that $w \models^+ B$ for all $B \in \Sigma$, $w \models^+ A$ also holds. Then, based on Theorem 5 and the completeness result for **N4** with respect to Kripke models (cf. Proposition of [18, p.425]) we also have:

---

[11] The readers may wonder if we may define the semantic clause $w \models^- \circ A$ as the equivalence: $w \models^+ A$ iff $w \models^- A$. However, this semantic clause will break the persistency requirement. This is one reason why we employ the current version of the semantic clause.

Proposition 3 *Given any set $\Sigma \cup \{A\}$ of formulas, $\Sigma \models_{\mathbf{c}} A$ iff $\Sigma \vdash_{\mathbf{cBS4}} A$.*

With the help of soundness of **cBS4**, we obtain the following.

Proposition 4 *Neither $\circ \circ A$ nor $A \vee (A \supset B)$ is derivable in* **cBS4**.

Proof Define $W := \{a, b\}$, $\leq := \{(a, a), (a, b), (b, b)\}$, $V^+(p) := \{b\}$ and $V^-(p) := \emptyset$. Then, a Kripke model $(W, \leq, V)$ satisfies persistency. We first show that $a \not\models^+ \circ \circ p$. Since $b \models^+ p$ and $b \not\models^- p$, we obtain $a \not\models^- \circ p$. By $a \not\models^+ p$ and $a \not\models^- p$, we obtain $a \not\models^+ \circ p$. Since we have established $a \not\models^- \circ p$ and $a \not\models^+ \circ p$, we can obtain $a \not\models^+ \circ \circ p$, which implies $\nvdash_{\mathbf{cBS4}} \circ \circ p$, as desired. As for $A \vee (A \supset B)$, we reuse the same $(W, \leq)$. Define $V^+(p) := \{b\}$ and $V^-(p) = V^+(q) = V^-(q) := \emptyset$. Then, $a \not\models^+ p$. Since $b \models^+ p$ and $b \not\models^+ p$, $a \not\models^+ p \supset q$. Thus $a \not\models^+ p \vee (p \supset q)$, as required. □

Proposition 4 suggests that, in order to obtain a natural deduction system corresponding to **cBS4**, we need to drop not only $(\supset_1)$ but also $(\circ_5)$ from $\mathcal{N}_2$**BS4**. However, we do not know if the resulting system is equipollent with **cBS4**. On the other hand, if we drop $(\supset_1)$ from $\mathcal{N}_1$**BS4** (we refer to this system as $\mathcal{N}$**cBS4**), we can easily see that the resulting calculus is equipollent with **cBS4**. Based on these considerations, we can also show that the classicality operator $\circ$ of **cBS4** can be employed without the intuitionistic conditional $\supset$. A key idea is to reformulate the rules $(\circ \supset_1)$, $(\circ \supset_2)$, and $(\sim \circ \supset)$ of Table 2 containing both $\circ$ and $\supset$ into the rules without $\supset$ as follows.

Definition 14 $\mathcal{N}$**cBD**$\circ$ consists of $(\circ_1)$, $(\circ_2)$, $(\circ_3)$, $(\circ_4)$, $(\mathbf{c}\circ_1)$, $(\mathbf{c}\circ_2)$, and $(\mathbf{c} \sim \circ)$ of Table 2 in addition to all the rules of $\mathcal{N}$**BD**.

Theorem 6 *Given any set $\Sigma \cup \{A\}$ of formulas, $\Sigma \vdash_{\mathcal{N}\mathbf{cBD}\circ} A$ iff $\Sigma \models_{\mathbf{c}} A$.*

Proof (Outline) Here, we concentrate on the completeness direction, i.e., the right-to-left direction. We show the contrapositive implication, so suppose $\Sigma \nvdash_{\mathcal{N}\mathbf{cBD}\circ} A$. Similarly as before in the completeness proof of $\mathcal{N}$**BD**$\circ$, we can find a *prime theory* Σ^+ (with respect to $\mathcal{N}$**cBD**$\circ$) such that $\Sigma \subseteq \Sigma^+$ and $A \notin \Sigma^+$. Let us define the canonical Kripke model as follows: W is all the non-trivial prime theories (we say that Γ is *non-trivial* if $B \notin \Gamma$ for some B), $V^+(p) = \{\Gamma \in W \mid p \in \Gamma\}$ and $V^-(p) = \{\Gamma \in W \mid \sim p \in \Gamma\}$ ($p \in \mathsf{Var}$). Then, $(W, \subseteq, V)$ is a Kripke model. Now we can establish the following equivalences: $\Gamma \models^+ C$ iff $C \in \Gamma$ and $\Gamma \models^- C$ iff $\sim C \in \Gamma$. Let us focus on the case where C is of the form of $\circ B$. We can establish the positive clause by $(\circ_1)$, $(\circ_2)$, $(\mathbf{c}\circ_1)$, $(\mathbf{c}\circ_2)$ and the negative clause by $(\circ_3)$, $(\circ_4)$, and $(\mathbf{c} \sim \circ)$. Then, we have $\Sigma^+ \models^+ C$ for all $C \in \Sigma$ but $\Sigma^+ \models^+ A$, which implies $\Sigma \not\models_{\mathbf{c}} A$, as required. □

6 Conclusion

Our original motivation was to introduce and generalize the notion of classicality in **BS4** to related systems such as those of Belnap-Dunn and Nelson. But it is not obvious at all if the classicality operators in these three systems have any common features. So, let us briefly consider this question from semantic and proof-theoretic perspectives before closing the paper. For this purpose, note that the relation between **BS4** and **cBS4** is just like classical logic and intuitionistic logic. And in such a case, the latter provides the uniform perspective, and the former will be captured as a degenerated case. This also applies to the following discussion.

Now, from the semantic viewpoint, what we kept in considering the semantics of $\mathscr{N}_2$**BS4** and $\mathscr{N}$**BD**$\circ$ is the truth table for the consistency operator $\circ$. But the truth table can be reformulated into a pair of conditions that reflect relational semantics provided by Dunn as follows[12]:

$$1 \in v(\circ A) \text{ iff } (1 \in v(A) \text{ and } 0 \notin v(A)) \text{ or } (1 \notin v(A) \text{ and } 0 \in v(A))$$
$$0 \in v(\circ A) \text{ iff } (1 \in v(A) \text{ iff } 0 \in v(A))$$

And comparing this with the semantic conditions for $\circ$ in **cBS4**, we can see that the above conditions are the degenerate cases of those for $\circ$ in **cBS4**. Thus, we may conclude that the classicality operator we dealt with in this paper can be characterized by the semantic conditions for $\circ$ in **cBS4**. Note here that (1) can be seen as the proof-theoretical representation reflecting the first condition above.

On the other hand, from the proof-theoretic viewpoint, we worked with both Hilbert-style systems and natural deduction systems. And for the basic observations, we employed the former to see the connection with the existing results, whereas we made use of the latter for considering the two question. The key was to provide two kinds of natural deduction system $\mathscr{N}_1$**BS4** and $\mathscr{N}_2$**BS4** for **BS4** which was the base system of our work. And the systems $\mathscr{N}$**BD**$\circ$ and $\mathscr{N}$**cBS4** that answer to our questions have common rules related to $\circ$. Those are ($\circ_1$), ($\circ_2$), ($\circ_3$) and ($\circ_4$). In view of the semantic considerations, these rules reflect only the left-to-right directions of the semantic conditions for $\circ$ in **cBS4**, and the rules corresponding to the other way around are not included as common rules. This is due to the lack of expressive power of natural deduction compared to sequent calculus, and we may obtain the corresponding proof-theoretical characterization through sequent calculus. Details will be kept for another occasion.

In this way, we may summarized our understanding of the classicality operator considered in this paper by the semantic conditions for $\circ$ in **cBS4**. But, this is only a tentative characterization of the notion of classicality, and we hope to find a better characterization that captures the notion in a more wide context by further investigations.

Finally, as we noted in the beginning of the paper, paraconsistent logicians disagree on many points. The only point on which they agree is in distinguishing theories being

[12] We already made use of this kind of reformulation in the proof of Theorem 3.

contradictory and trivial. However, there seems to be another implicit agreement. That is, many of the paraconsistent logicians do not necessarily abandon classical logic completely. Indeed, in consistent (and complete) cases, they do agree that classical logic works well.[13] And in view of this fact, it seems that the idea of classicality operator is acceptable by many of the paraconsistent logicians. But, in the literature, the notion of classicality has been always introduced together with the material conditional and therefore its wide applicability seems to have been not recognized. Our hope is that readers now have a different impression. Needless to say, however, our result is just a first step, as we have only examined two special cases. Many questions are left open. Those include investigations into various theories, such as naive set theory and naive truth theory, based on systems with a classicality operator, and philosophical justification of the notion of classicality. We shall leave these topics, together with others mentioned in the paper earlier, for another occasion.

Acknowledgments The authors would like to thank the referees for their detailed and helpful comments which improved the paper in many ways. We would also like to thank Michael De who kindly proofread our final draft and made many helpful suggestions to improve our remarks related to relevantists' perspective as well as our English. Finally, we would like to thank the audiences at the Trends in Logic XI conference who showed their interest, and encouraged us to pursue this research. The first author is a postdoctoral fellow of Japan Society for the Promotion of Science (JSPS), and the present work was partially supported by a Grant-in-Aid for JSPS Fellows. The work of the second author was partially supported by JSPS KAKENHI, Grant-in-Aid for Young Scientists (B) 24700146.

References

1. Belnap, N. (1976). How a computer should think. In G. Ryle (Ed.), *Contemporary aspects of philosophy* (pp. 30–55). Stocksfield: Oriel Press.
2. Carnielli, W. A., Coniglio, M. E., & Marcos, J. (2007). Logics of formal inconsistency. In D. Gabbay & F. Guenthner (Eds.), *Handbook of philosophical logic* (Vol. 14, pp. 1–93). Dordrecht: Springer.
3. Carnielli, W. A., & Marcos, J. (2002). A taxonomy of C-systems. In W. A. Carnielli, M. E. Coniglio & d'Ottaviano I. M. L. Marcel Dekker (Eds.), *Paraconsistency: The Logical Way to the Inconsistent, Proceedings of the II World Congress on Paraconsistency* (pp. 1–94).
4. Carnielli, W. A., Marcos, J., & de Amo, S. (2000). Formal inconsistency and evolutionary databases. *Logic and Logical Philosophy*, *8*, 115–152.
5. de Castro, M. A., & d'Ottaviano, I. M. L. (2000). Natural deduction for paraconsistent logic. *Logica Trianguli*, *4*, 3–24.
6. da Costa, N. C. A. (1974). On the theory of inconsistent formal systems. *Notre Dame Journal of Formal Logic*, *15*(4), 497–510.
7. Guillaume, M. (2007). Da Costa 1964 logical seminar: Revisited memories. In J.-Y. Béziau, W. A. Carnielli & D. Gabbay (Eds.), *Handbook of paraconsistency* (pp. 3–62). London: College Publications.

[13] Note that even though many of the relevant (relevance) logics are paraconsistent, relevantists have a different view on classical logic in the sense that they will not necessarily agree to make use of classical logic even in consistent cases.

8. Jaśkowski, S. (1999). A propositional calculus for inconsistent deductive systems. *Logic and Logical Philosophy*, *7*, 35–56.
9. Kamide, N., & Wansing, H. (2012). Proof theory of Nelson's paraconsistent logic: A uniform perspective. *Theoretical Computer Science*, *415*, 1–38.
10. Loparić, A., & da Costa, N. C. A. (1984). Paraconsistency, paracompleteness, and valuations. *Logique et Analyse (N.S.)*, *106*, 119–131.
11. Mendelson, E. (1997). *Introduction to mathematical logic*. Boca Raton: Chapman and Hall/CRC.
12. Odintsov, S. P. (2008). *Constructive negations and paraconsistency*. Dordrecht: Springer.
13. Omori, H., & Waragai, T. (2011). Some observations on the systems LFI1 and LFI1*. In: *Proceedings of Twenty-Second International Workshop on Database and Expert Systems Applications (DEXA2011)* (pp. 320–324).
14. Priest, G. (1979). The logic of paradox. *Journal of Philosophical Logic*, *8*, 219–241.
15. Priest, G. (2002). Paraconsistent logic. In D. Gabbay & F. Guenthner (Eds.), *Handbook of philosophical logic* (2nd ed., Vol. 6, pp. 287–393). Dordrecht: Kluwer Academic Publishers.
16. Sano, K., & Omori, H. (2013). An expansion of First-order Belnap-Dunn logic. *Logic Journal of IGPL*, doi:10.1093/jigpal/jzt044.
17. Sette, A. (1973). On the propositional calculus P^1. *Mathematica Japonicae*, *16*, 173–80.
18. Wansing, H. (2001). Negation. In L. Goble (Ed.), *The Blackwell guide to philosophical logic* (pp. 415–436). Cambridge: Blackwell Publishing.

Explicating the Notion of Truth Within Transparent Intensional Logic

Jiří Raclavský

Abstract The approach of Transparent Intensional Logic to truth differs significantly from rivalling approaches. The notion of truth is explicated by a three-level system of notions whereas the upper-level notions depend on the lower-level ones. Truth of possible world propositions lies in the bottom. Truth of hyperintensional entities—called constructions—which determine propositions is dependent on it. Truth of expressions depends on truth of their meanings; the meanings are explicated as constructions. The approach thus adopts a particular hyperintensional theory of meanings; truth of extralinguistic items is taken as primary. Truth of expressions is also dependent, either explicitly or implicitly, on language (its notion is thus also explicated within the approach). On each level, strong and weak variants of the notions are distinguished because the approach employs the Principle of Bivalence which adopts partiality. Since the formation of functions and constructions is non-circular, the system is framed within a ramified type theory having foundations in simple theory of types. The explication is immune to all forms of the Liar paradox. The definitions of notions of truth provided here are derivation rules of Pavel Tichý's system of deduction.

Keywords Truth · Truth of propositions · Truth of expressions · Language · Transparent intensional logic

1 Introduction

I suggest an explication of the notion *true* within the extensive logical framework of Pavel Tichý's *Transparent Intensional Logic* (*TIL*). The approach differs significantly from other well-known approaches to truth such as the hierarchical and bivalent proposal by Tarski (1933/1956), three-valued theories by Kripke [7] and others,

J. Raclavský (✉)
Department of Philosophy, Masaryk University, Brno, Czech Republic
e-mail: raclavsky@phil.muni.cz

R. Ciuni et al. (eds.), *Recent Trends in Philosophical Logic*,
Trends in Logic 41, DOI: 10.1007/978-3-319-06080-4_12,

paraconsistent dialetheism by Priest [8], revision theory by Gupta and Belnap [4], paracompleteness by Field [3] and Beall [1], axiomatic approaches by Halbach [5] and Horsten [6], etc. The brevity of space does not allow to provide a comparison of the present approach with the aforementioned ones; nevertheless, some differences can be read from the rest of this introduction and some other remarks in the paper.

The key feature of the present approach is that the truth of certain non-linguistic entities is construed as primary, while the truth of linguistic entities, which represent the non-linguistic ones, is defined as dependent on it.

The notion of truth, as explicated in TIL, splits in three kinds according to the range of their applicability to:

a. propositions (which can be considered to be denotata of expressions),
b. (so-called) constructions of propositions (which can be considered to be meanings of expressions),
c. expressions.[1]

The notions of the kinds a. and b. are obviously independent on language and precede the notion of the kind c.

Truth of propositions—where possible world propositions are classes of world-time couples—is rather transparent: a proposition is true in a given possible world W at a moment of time T iff its value for this $\langle W, T\rangle$ is the truth-value T. Then, truth of constructions is best definable in terms of truth of propositions constructed by them.[2]

Constructions are abstract structured entities akin to algorithms; they construct objects, e.g. propositions. Constructions are 'intensional' entities, thus they can aptly serve for the recently urged hyperintensional individuation of meanings.

The notions of the kind c., truths of expressions, are dependent on, and relative to language(s). The relativity is either explicit, or implicit. Truth of expressions is defined in terms of truth of the expressions' meanings (denotata). Thus unlike the usual approach of Tarski and others, the proposed explication does not depend on the notion of translation (recall that Tarski's method requires that an expression is translated to the theoretician's metalanguage). On the other hand, the present approach relies on the (explicated) notion of language.

It can be shown that the explication resists all forms of the Liar paradox. The explication also confirms Tarski's famous Undefinability theorem, though in a bit supplemented form.

[1] Such gradual construction was in fact suggested by Tichý in his remarks on truth [13, Chaps. 11 and 12]. There, certain (verbal) definitions of the notions can be found. Tichý's investigations surely inspired my approach. The present paper is an extract from a large manuscript on truth; some of my results have been published in Raclavský [10].

[2] It is in the spirit of intensional explication of our conceptual scheme to say that propositions can be construed as facts and our world can be construed as a collection of (actual) facts. Then, the proposal of TIL confirms a sort of correspondence theory of truth (true sentences correspond to facts that obtain). However, these issues cannot be discussed here.

It is also important to stress that the present approach is in some important sense neo-classical. Classical rules, including the Principle of Bivalence,[3] are preserved. Because of partiality adopted in the system, however, the rules are appropriately modified. It has a certain connection with the fact that, for each level of truth-notions, there are distinguished total (strong) and partial (weak) variants of the notions.

Employing the truth of non-linguistic entities (propositions and constructions), the approach is immune to well-known arguments of the philosophy of language against 'linguistic' treatment of semantic matters. Moreover, the explication of truth by TIL relies on a hyperintensional (procedural) way of explication of meanings.

TIL is based on λ-calculus accompanied by a particular ramified theory of types. It means that it is a very expressive language within which various axiomatic theories (systems) can be formulated (it is thus not an aim of this paper to state any such particular theory or system, cf. also below).

Unfortunately, the lack of space does not enable us to discuss any such matter in greater detail. Moreover, an explication of various particular notions of truth which might come to one's mind cannot be provided here, although the approach is capable of such explication.

The paper is organized as follows. The Sect. 2 explains briefly the notion of construction, deduction, type theory, and explication of meanings. The Sects. 3 and 4 suggest explications of the two kinds of language-independent notions of truth, which are mentioned in titles. The penultimate Sect. 5 begins with an explication of language, which is needed especially for the explication of truth of expressions which are explicitly relative to language. Then, truth of expressions which is implicitly relative to language is explicated and the resistance to the Liar paradox is shown. Finally, the limitation of language and thus also the Undefinability Theorem will be briefly discussed.

2 Elements of TIL

The basic ideas of TIL will best be introduced by the following, partly historical, story. In the late 1960s, Tichý began to utilize Church's simple theory of types (i.e. typed λ-calculus) for logical analysis of natural language. To its basic sorts (atomic types) of *individuals* and *truth-values* (T and F), Tichý added two other sorts—those of *possible worlds* and *moments of times/real numbers*.[4] Together with some

---

[3] The *Principle of Bivalence* adopted here reads as: for any proposition P, P has at most one of the two truth-values T and F in a given W and T. In other words, a proposition can be gappy; for instance, the proposition "The king of France is bald" is gappy in the actual W and present T. (Note that I use single quotation marks for quotation of expressions or, sometimes, for indication of a shift in meaning; double quotation marks are used for indication of propositions and other extralinguistic entities.)

[4] In TIL, possible world *intensions* (i.e. *propositions*, *properties*, *relations-in-intension*, etc.) are total or partial functions from world-time couples to certain entities (viz. truth-values, classes of objects, classes of n-tuples of objects, etc.). Among *non-intensions* one can find in TIL, e.g., the

semantic doctrines concerning ways to analyze the meaning of an expression, the framework began to rival the much more popular system of Montague. Tichý also soon adopted partiality and he mainly explicated a number of phenomena associated with meaning: modalities, propositional attitudes, intensional transitives, descriptions, temporal adjectives, verb tenses, verb aspects, etc.

The second important feature of TIL are its hyperintensional entities. In early 1970s, Tichý realized that possible world intensions are too coarse-grained to be proper meanings of expressions; rather, one needs structured *hyperintensions*.[5] Two main kinds of λ-terms are usually understood as denoting values of functions or functions as such, but Tichý noticed that they can be also understood as expressing applications of functions to arguments or ways of obtaining functions. On the latter, 'intensional', reading of λ-terms, these stand for constructions, i.e. TIL's hyperintensional entities. Some constructions might also be understood as functions in the older sense, i.e. functions as procedures (rules), which contrasts with the modern notion of function as a mere mapping.

Constructions are procedural entities, akin to algorithmic computations (they are not purely set-theoretical objects). Constructions are language independent; TIL λ-terms serve only to depict constructions (in other words, the formal language of TIL has fixed interpretation). Each object, e.g. a proposition, is constructed by infinitely many equivalent but not identical constructions (constructions thus satisfy intensional principle of individuation). Each construction C is specified by two things: i. the object O constructed by C, ii. the way C constructs, dependently on valuation v, the object O (by means of which subconstructions). Note that constructions are closely connected with objects constructed by them.

For a defence of the notion of construction showing mainly its need, cf. especially Tichý's book [13]. For the application of TIL to natural language analysis, see Tichý [14], Duží et al. [2], or Raclavský [10]. All these books also include various other applications of TIL. For the rest of the paper, we need to bear in mind at least the following matters concerning semantic scheme, specification of constructions, type theory and deduction (consult the aforementioned books for technical details).

In order to explicate meanings of (natural) language, Tichý employed a *semantic scheme* which can be précised as follows:

an expression E	
| *E expresses* (*means*) in L:	
a construction C	= the *meaning* of E in L
| *C constructs*:	
an intension/non-intension	= the *denotatum* of E in L

(Footnote 4 continued)
well-known classical truth-functions $\neg$, $\rightarrow$, $\wedge$, and $\vee$, the well-known subclasses of classes of ξ-objects $\exists^{\xi}$ and $\forall^{\xi}$ (for any type ξ; the indication '$^{\xi}$' will be usually suppressed), or the well-known identity relation between ξ-objects, $=^{\xi}$. Constructions are also non-intensions.

[5] One of the notorious arguments for adoption of hyperintensions is that due to intensional analysis, beliefs which are equivalent but non-identical are merged to one. On such use of possible world propositions, an argument that one believes that $1+1=2$ thus one believes Fermat's Last Theorem is wrongly rendered as valid.

Empirical expressions ('the Pope', 'tiger', 'It rains in Nice', ...) denote intensions; non-empirical expressions ('not', '3', ...) denote non-intensions. The value of an intension in W at T is the *referent* in L, W and T of an empirical expression. The denotatum in L and referent in L, W and T of a non-empirical expression are construed as identical.

Constructions divide into six kinds according to the ways of their constructing. Let X be any object or construction and C_i be any construction (of order k):

i. *variable* x_k v-constructs the k-th object (of an appropriate type) of the valuation v;
ii. *trivialization* ${}^0X$ v-constructs (for any v) the object X directly, without any change (${}^0X$ takes X and leave it as it is);
iii. *single execution* ${}^1X$ v-constructs the object (if any) v-constructed by X;
iv. *double execution* ${}^2X$ v-constructs the object (if any) which is v-constructed by the construction (if any) v-constructed by X;
v. *composition* $[C \;\; C_1 \ldots C_n]$ v-constructs the value (if any) of the function F (if any) v-constructed by C on the string of entities $A_1 \ldots A_n$ (if any) v-constructed by $C_1, \ldots, C_n$;
vi. *closure* $\lambda x C$ v-constructs (for any v) a function which maps the objects in the range of x to the objects which are v-constructed by C (a very much simplified formulation).

Note that the constructions of the kinds iii–v can be abortive in the sense that they v-construct nothing whatsoever, they are *v-improper* constructions. For instance, a composition is v-improper when the partial function v-constructed by C is not defined on the string of entities v-constructed by $C_1, \ldots, C_n$. Two constructions are *v-congruent* iff they v-construct one and the same object or they are both v-improper.

The lack of space does not enable me to repeat here Tichý's whole definition [13, p. 66] of his unique *ramified type theory*. In the basis of the hierarchy, there are atomic types. In case of TIL, for instance, these are types of individuals, truth-values, possible worlds, and moments of times. The rest of first-order types cover all total and partial n-ary functions over the objects belonging to the first-order types (i.e. first-order objects). Higher-order types include especially types for constructions. For instance, there is a particular type containing the k-order constructions, i.e. constructions of the k-order objects (for $1 \leq k \leq n$). Moreover, functions from or to constructions are classified by some higher-order types as well. It is readily seen that the hierarchy of entities is very, very rich.[6]

In several of his papers (cf. [14]), Tichý also exposed a *deduction system* for his type theory, thus also for TIL. Its derivation rules are made from sequents whereas sequents are made from so-called matches; matches consist of constructions and (trivializations of) objects v-constructed by them. Sequents and rules are thus not expressions of a formal language. Derivation rules display properties of objects. To

[6] The stratification of entities into such hierarchy is justified by *four Vicious Circle Principles* [10], each of them being entailed by the *Principle of Specification*: you cannot fully specify an entity by means of the entity itself.

illustrate, the derivation rule $\Phi \cup \{{}^0\mathrm{T} : o_1\} \Rightarrow {}^0\mathrm{T} : o_2 \models \Phi \Rightarrow {}^0\mathrm{T} : [o_1 \; {}^0\!\!\rightarrow o_2]$, where o_1 and o_2 are variables for truth-values, shows that the material conditional $\rightarrow$ returns T for the couple $\langle \mathrm{T}, \mathrm{T}\rangle$.[7]

Constructions and derivation rules can be organized in *derivation systems* [9]. Roughly speaking, they are objectual correlates of axiomatic systems. It follows from the very notion of derivation system that no derivation system can be separated from its objectual area. Thus, if one has (say) a property of truth at one's disposal, it is not inevitable for one to build up a particular derivation system to single out which particular object is the truth property in question. A derivation system is rather a tool for proving facts about an object (say the truth property), whereas the facts are implied by features of the object.

3 Truth of Propositions

Truth of propositions is a phenomenon dependent on circumstances, i.e. possible worlds and moments of time. For a proposition to be true in W at T is nothing but having T as a value for that world-time couple.

The notion splits in two variants: the partial and the total one. According to the *partial notion*, a proposition P which is gappy in a given W and T is not true or false. In this case—i.e. when the valuation v assigns such P to the variable p, W to the variable w, and T to the variable t—the construction p_{wt},[8] and thus also $[{}^0\mathrm{True}^{\pi\mathrm{P}}_{wt}\, p]$, is v-improper:

$$[{}^0\mathrm{True}^{\pi\mathrm{P}}_{wt}\, p] \Leftrightarrow^{o} p_{wt} (\text{alternatively } [p_{wt} \; {}^0\!\!= {}^0\mathrm{T}])$$

The extension of the property "True$^{\pi\mathrm{P}}$" in W at T, i.e. a characteristic function, is undefined for P. The definition matches the deflationist intuition that there is a notion of truth which adds nothing to a proposition.

According to the *total notion*, on the other hand, a proposition which is gappy in a given W and T is assigned by the truth-value F (the variable o ranges over the type of truth-values):

$$[{}^0\mathrm{True}^{\pi\mathrm{T}}_{wt}\, p] \Leftrightarrow^{o} [{}^0\exists\, \lambda o[[o \; {}^0\!\!= p_{wt}] \; {}^0\!\!\wedge [o \; {}^0\!\!= {}^0\mathrm{T}]]]$$

---

[7] I view *definitions* as certain $\Leftrightarrow$-rules (both $\Rightarrow$ and $\Leftrightarrow$ concern satisfiability of sequents). Two constructions flanking $\Leftrightarrow^{\xi}$ are v-congruent for any v; the type of the object v-constructed by both constructions will be indicated nearby '$\Leftrightarrow$'. Definitions can also be viewed as proposing an explication of the intuitive notion whose rigorous correlate occurs in the left hand side of the definition; its right hand side shows in which sense the notion 'is meant', which objects 'fall under' it, cf. [10].

[8] 'C_{wt}' abbreviates '$[[C\; w]\; t]$'.

In those W and T, a gappy proposition thus falls in the antiextension of the property "True$^{\pi\mathrm{T}}$". Every proposition is thus determinately assessed as true or not true.[9]

4 Truth of Constructions

The notions of truth of constructions form two groups. In the first group, there are notions of truth *independent on circumstances*. A sample verbal definition: a (k-order) construction is true (or rather: is a *truth*) iff it (v-)constructs the truth-value T. In the definition, the variable c^k ranges over the type of k-order constructions and the double execution corresponds to the word '(v-)constructs':

$$[{}^0\mathrm{Truth}^{*k\mathrm{P}}c^k] \Leftrightarrow^{\mathrm{o}} [{}^2c^k \; {}^0{=} \; {}^0\mathrm{T}]$$

L-truths can be defined as truths v-constructing T on every v.[10]

In the second group, there are kinds of truths of constructions which are *dependent on circumstances*. The sensitivity on circumstances can be traced back to the circumstance sensitivity of truth of propositions which are (v-)constructed by the constructions.

Of course, there is a plenitude of such particular (sub)kinds of truths of constructions because of the plenitude of orders of constructions. (A hierarchy.) But there are also various distinct notions of truth of constructions within one particular order, which corresponds to the fact that there are various slightly distinct notions of truth.[11]

For instance, we have both partial and total notions of truth of constructions. In the definition, ${}^2c^k_{wt}$ v-constructs the value (if any) of the proposition (if any) v-constructed by the construction (if any) v-constructed by c^k:

$$[{}^0\mathrm{True}^{*k\mathrm{P}}_{wt}c^k] \Leftrightarrow^{\mathrm{o}} [{}^0\mathrm{True}^{\pi\mathrm{P}}_{wt} \; {}^2c^k]$$

$$[{}^0\mathrm{True}^{*k\mathrm{T}}_{wt}c^k] \Leftrightarrow^{\mathrm{o}} [{}^0\exists \; \lambda o[[o \; {}^0{=} \; {}^2c^k_{wt}] \; {}^0{\wedge} \; [o \; {}^0{=} \; {}^0\mathrm{T}]]]$$

But there are even other notions. To give at least one example from a range of several similar notions definable within the framework, let us define a notion according to which constructions of propositions are only determinately assessed as true or not true, while all other constructions (of individuals, of classes of numbers, ...) are left unassessed:

$$[{}^0\mathrm{True}^{*k\mathrm{PT}}_{wt}c^k] \Leftrightarrow^{\mathrm{o}} [{}^0\mathrm{True}^{\pi\mathrm{T}}_{wt} \; {}^2c^k]$$

---

[9] It is just this notion which should be deployed in appropriate reformulations of classical laws in order to be valid within a framework adopting partiality.

[10] For that purpose a bit richer type basis is needed.

[11] Some of them might be defined also by other theoreticians (assuming here translatability of their results to the present framework).

As regards the constructing of the definiens, if ${}^2c^k$ does not v-construct a proposition, the actual extension of the property "True$^{\pi\mathrm{PT}}$" cannot be applied, thus the definiens is a v-improper construction (so is the definiendum). The classes (i.e. characteristic functions) which are extensions of the property "True$^{*k\mathrm{PT}}$" are thus partial.

5 Truth of Expressions

Truth of expressions is dependent on language(s).[12] It is thus intuitively correct to say that an expression E is true in L (in W at T) iff it is true (in W at T) what the expression E means in L. My explication matches this natural definition. The truth of expressions is apparently a semantic property or rather relation(-in-intension). In order to explicate the relation, the explication of the notion of language thus has to be undertaken.

Language is a normative system enabling speakers to communicate. It seems sufficient for our purposes to restrict our attention to the expressive, coding aspect of language in the synchronic sense and model it simply as a function from expressions to meanings. In TIL, a *k-order code* L^k is a (partial) function from (Gödelized) expressions to k-order constructions [13, p. 228].

But language such as English would be better modelled rather as a hierarchy of codes L^1, L^2, ..., L^n [10]. It corresponds to the existence of 'commenting', 'reflective' levels in language—language enables us to comment on its own parts. On such construal, most of everyday communication takes place in the first-order code of the hierarchy; higher-order coding means, which are used for commenting, are not frequently utilized. A particular hierarchy of codes is a class such that i. it involves n codes of n mutually distinct orders, ii. each expression E having a meaning M in L^k has the same meaning M in L^{k+m} $(1 < m)$, and iii. an expression E lacking meaning in L^k can be meaningful in L^{k+m}. One naturally adds also iv. *compositionality* within the codes of the hierarchy.

In consequence of this, every code of a particular hierarchy shares the same expressions as any other code of the same hierarchy; quantification over all of them is unrestricted. Due to order-cumulativity of functions, every k-order code is also a k+1-order code; the type involving n-order codes thus includes nearly all codes of the hierarchy; we can quantify over them. A hierarchy of codes is a certain class; such classes form an n-order type and we can thus quantify over them.

However, every code is limited in its expressive power because no construction of a k-order code L^k is codable in L^k, only in a higher-order code. Moreover, no expression mentioning (precisely: referring to) L^k is endowed with meaning in L^k, only in a higher-order code. To illustrate it, consider the immediate construction of L^k, viz. ${}^0L^k$, which is the meaning of 'L^k' (an expression referring to L^k). If ${}^0L^1$ were a value of L^1, L^1 would not be specifiable.

---

[12] Truth of expressions' tokens can be defined as dependent on truth of expressions. It is entirely omitted in this paper.

There are two groups of notions of truth of expressions. Let us begin with the first group. Each of its notions determines a relation(-in-intension) between expressions and languages (codes).[13] They are *explicitly language relative notions.*

A prototypical example can be defined as follows, note the perfect match of the definition with the intuitive claim stated in the beginning of this section (e is a variable for numbers/expressions, l^n is a variable for n-order codes):

$$[{}^0\mathrm{TrueIn}^{\mathrm{PT}}_{wt}\, e\ l^n] \Leftrightarrow^{\mathrm{o}} [{}^0\mathrm{True}^{*n\mathrm{PT}}_{wt}[l^n\ e]]$$

The defined notion is such that only expressions denoting (in the respective language) propositions are assessed as true or not true. This is not achieved in the case of the total notion of truth of expressions—its definition is not difficult to come by—which renders also all other expressions as not true.

Note that both the definiendum and the definiens are $n{+}1$-order constructions.[14] Hence, they cannot be expressed already in an n-order code (the point can be surely generalized also for $n = 1$). This is the reason why an appropriate version of the Liar paradox is avoided (cf. [11]).

Each notion from the second group determines a semantic property of expressions, not a relation(-in-intension). Unlike the preceding case, these notions are *implicitly language relative notions* of truth of expressions. Let us consider an example of a concrete definition:[15]

$$[{}^0\mathrm{True}^{\mathrm{L}n\mathrm{T}}_{wt}\, e] \Leftrightarrow^{\mathrm{o}} [{}^0\mathrm{TrueIn}^{\mathrm{T}}_{wt}\, e\ {}^0\mathrm{L}^n]$$

The definiens removes the ambiguity of the intuitive notion in question. It is thus quite clear that it is L^n rather than $\mathrm{L'}^n$ (belonging to the hierarchy of, say, German), or L^n rather than L^{n-1}, in which the semantic feature of an expression E should be examined. It is perhaps just this ambiguity ubiquitously present in our ordinary and even scientific thinking which is the source of the Liar paradox.

Unlike the definiens, which is an $n+1$-order construction, the definiendum already is, due to the order-cumulativity of constructions, an n-order construction. It may then seem that a revenge of the Liar paradox is possible. Can be the k-order construction (involving the total notion of truth of expression in L^k)

$$\lambda w \lambda t \lambda e[{}^0\neg\ [{}^0\mathrm{True}^{Lk\mathrm{T}}_{wt}\ e]]$$

expressed already in the k-order code L^k?

---

[13] To ask for an expression's meaning in a hierarchy of codes amounts to ask for its meaning in the (virtually) highest code of the hierarchy, i.e. L^n.

[14] The typing technique within Tichý's type theory is similar to that in Russellian ramified type theories.

[15] Recall that if properly closed by lambdas, both constructions flanking $\Leftrightarrow^{\mathrm{o}}$ v-construct one and the same property. It can be proved that the property cannot be discussed by L^n, cf. our discussion below.

The Tarskian approach to this question would utilize the appropriate version of the Liar paradox as a *proof* of the negative answer.[16] The negative answer can rely also on the intuitively valid fact—entailed, inter alia, by the proof based on the Liar paradox—that *no code with a sufficient expressive power enables us to discuss its own semantic features*.[17]

A very similar fact was concluded already by Tarski (1933/1956, the *Undefinability Theorem* [12]) and also by Tichý [13, p. 231, 233]. As regards the differences, note that on the present approach the lower-order semantic notions (seemingly expressible in object language) are definable, yet they are not expressible in the lower-order codes (object languages). In other words, the proper goal of logicians who investigate truth—viz. to construct a language with a truth-predicate applicable to the expressions of that very language—is not fully achievable if the language in question is sufficiently rich.[18]

6 Conclusions

To stress the essential feature of the TIL approach to truth, the notion of truth is explicated by a three-level system of notions. Truth of expressions is 'supervening', dependent on the lower-level notions of truth which apply to extralinguistic items serving as meanings/denotata of (some) expressions. The TIL approach thus differs significantly from rivalling approaches.

On each level, some novelties with regard to the present understanding are proposed. Truth of possible world propositions is a rather simple notion and it gives rise to no (semantic) paradoxes. Truth of constructions, Tichý's hyperintensional entities, splits into a number of variants along the regulations of the type theory which is governed by a special version of the Vicious Circle Principle. A Liar-like paradox does not ensue because of type restrictions. However, non-paradoxicality is not a primary goal of the implemented type ramification but a product of a reasonable formation of constructions. Truth of expressions depends on truth of constructions they express/propositions they denote. This is dependent, either explicitly or implicitly, on language; thus one has truth as a relation between expressions and languages and as a 'relational' property of expressions. The notion of language utilized here results in a hierarchy because of the type hierarchy of constructions. In consequence, the proposal is immune to all forms of the Liar paradox. Recall also that meanings are treated by this approach quite explicitly and that they are explicated as certain hyperintensional entities.

The approach may be seen as a certain 'neo-hierarchical' approach combining Russellian, Tarskian and Kripkean approaches. As regards Russell, however, only

[16] Cf. Tarski [12] and Tichý [13, pp. 292–293] or Raclavský [11] for such proofs.

[17] In insufficiently expressive codes-languages, a partial truth-predicate can be meaningful without a risk of the Liar paradox (cf. [11]).

[18] Sufficient richness was an original Tarski's condition, cf. his [12].

some Tichý's constructions roughly appear to be similar to the linguistic entities called 'propositional functions' by Russell. Thus, also the hierarchy of constructions only has an extraneous similarity to Russell's hierarchy of propositional functions. Further, note that Tichý's particular version of the ramified type theory has foundations in simple theory of types. Tarski's hierarchy of languages provides a much better example of a comparable proposal. But the essential difference is that the present hierarchy is based on the hierarchy of constructions which are meanings of expressions, while Tarski did not investigated meanings of the expressions belonging to the languages he considered. The analogy with Kripke can be retained if one ignores some important features of Kripke's proposal, maintaining only that truth comes in total and partial versions.

References

1. Beall, J. C. (2009). *Spandrels of truth*. Oxford: Oxford University Press.
2. Duží, M., Jespersen, B., & Materna, P., (2010). *Procedural semantics for hyperintensional logic*. New York: Springer.
3. Field, H. (2008). *Saving truth from paradox*. Oxford: Oxford University Press.
4. Gupta, A., & Belnap, N. (1993). *The revision theory of truth*. Cambridge: A Bradford Book.
5. Halbach, V. (2011). *Axiomatic theories of truth*. Cambridge: Cambridge University Press.
6. Horsten, L. (2011). *The Tarskian turn: deflationism and axiomatic truth*. Cambridge: The MIT Press.
7. Kripke, S. A. (1975). Outline of a theory of truth. *Journal of Philosophy*, *72*(19), 690–716.
8. Priest, G. (1987). *In contradiction*. Boston: Martinus Nijhoff Publishers.
9. Raclavský, J., & Kuchyňka, P. (2011). Conceptual and derivation systems. *Logic and Logical Philosophy*, *20*(1–2), 159–174.
10. Raclavský, J. (2009). *Names and descriptions: logico-semantical investigations*. Olomouc: Nakladatelství Olomouc. (In Czech).
11. Raclavský, J. (2012). Semantic paradoxes and transparent intensional logic. In M. Peliš & V. Punčochář (Eds.), *The Logica Yearbook 2011* (pp. 239–252). London: College Publications.
12. Tarski, A. (1933). The notion of truth in formalized languages. In: Logic, semantics and metamathematics. Oxford: Oxford University Press.
13. Tichý, P. (1988). *The foundations of Frege's logic*. Berlin: Walter de Gruyter.
14. Tichý, P. (2004). *Pavel Tichý's collected papers in logic and philosophy*. Dunedin: Filosofia .

Leibnizian Intensional Semantics for Syllogistic Reasoning

Robert van Rooij

Abstract Venn diagrams are standardly used to give a semantics for Syllogistic reasoning. This interpretation is *extensional*. Leibniz, however, preferred an *intensional* interpretation, according to which a singular and universal sentence is true iff the (meaning of) the predicate is *contained in* the (meaning of) the subject. Although Leibniz's preferred interpretation played a major role in his philosophy (in Leibniz [16] he justifies his metaphysical 'Principle of Sufficient Reason' in terms of it) he was not able to extend his succesfull intensional interpretation (making use of characteristic numbers) without negative terms to one where also negative terms are allowed. The goal of this paper is to show how syllogistic reasoning with complex terms can be given a natural set theoretic 'intensional' semantics, where the meaning of a term is not defined in terms of individuals. We will make use of the ideas behind van Fraassen's [6, 7] hyperintensional semantics to account for this.

Keywords Syllogisms · Leibniz · Intensional semantics · Negation

1 Introduction

Aritstotle made in his Prior Analytics a distinction between assertoric and modal syllogistics. The crucial difference between the two syllogistics is that only the latter makes use of two different types of predicative relations: accidental versus essential predication. 'Animal' is essentially predicated of 'mammal', but 'walking' is not. Although both (1) 'Every man walks' and (2) 'Every man is an animal' can be true, it is natural to say that the 'reasons' for their respective truths are different. Sentence (1) is true by accident, just because every actual man happens to (be able to) walk. The sentence (2), on the other hand, is true because manhood necessarily

R. van Rooij (✉)
Institute for Logic, Language and Computation, Universiteit van Amsterdam, Amsterdam, The Netherlands
e-mail: R.A.M.vanRooij@uva.nl

R. Ciuni et al. (eds.), *Recent Trends in Philosophical Logic*, Trends in Logic 41, DOI: 10.1007/978-3-319-06080-4_13,

involves being animate. In traditional terms it is said that (2) is true *by definition*, although this notion of 'definition' should not be thought of nominalistically: it is the *real* definition. A natural way to account for accidental predication is to say that a sentence of the form 'Every S is P' is true just in case every actual S-*individual* is also a P-*individual*. A natural way to account for essential predication, on the other hand, is to say that a sentence of the form 'Every S is P' is true just in case the *real* definition of S (the set of attributes one needs to have to be an S) includes the real definition of P (the set of attributes one needs to have to be a P). We will say that the first way to determine whether 'Every S is P' is true is *extensional* in nature, the second way *intensional*. Especially due to the influence of the Port-Royal school of logic, however, it became standard in the seventeenth century to assume that the two come down to the same thing. Leibniz explicitly endorsed this position. Leibniz gave an intensional semantics making use of characteristic numbers. Unfortunately, he was unable to extend this system with both conjunctive and negative terms. The main goal of this paper is to see how to make sense of the intensional interpretation of syllogistic reasoning including both type of terms.

2 Traditional Syllogistic Reasoning

A categorical sentence always contains two *terms*. A categorical sentence is always of one of four kinds: *a*-type: Universal and affirmative ('All men are mortal', with the terms 'men' and 'mortal'); *i*-type: Particular and affirmative ('Some men are philosophers'); *e*-type: Universal and negative ('No philosophers are rich'), and *o*-type: Particular and negative ('Some men are not philosophers'). Thus, the *syntax* of simple categorical sentences can be formulated as follows: If T and T' are terms, TaT', TiT', TeT', and ToT' are categorical sentences. Syllogisms are arguments in which a categorical sentence is derived as conclusion from two categorical sentences as premisses. A *valid* syllogism is a syllogism that cannot lead from true premisses to a false conclusion. It is well-known that with the help of Venn-diagrams one can check which syllogisms are valid (cf. [4]). For some, its validity depends on whether or not we assume existential import. The traditional (proof-theoretic) way to check validity, however, was to see whether the syllogism could be reduced to the first four valid syllogisms of the first figure (Barbara, Celarent, Darii, and Ferio) with the help of conversion and *reductio ad impossible*.

The syntax of categorical sentences can be straightforwardly extended with conjunctive and negative terms.[1] Thus we say that if T and T' are terms, $\overline{T}$ and TT' are terms as well. In terms of Venn-diagrams it is still easy to see which syllogisms are valid. Still, negative terms didn't play an important role in traditional logic. Arguably, this was no accident, and due to the fact that an interpretation in terms of Venn-diagrams was foreign to traditional logicians. Making use of Venn-diagrams assumes that terms are interpreted *extensionally*: as sets of individuals. According to

[1] In the history of logic, negative terms are also known as *indefinite* or *infinite* terms.

alternative interpretations, however, a sentence of the form '*SaP*' is true if and only if the *essence*, *comprehension*, or *intension*, of *P* is contained in the *intension* of *S*. The *intension* of a term, or concept, consisted of all the essential attributes in it (those that cannot be removed without 'destroying' the concept). Thus, the intension of the term 'triangle' might include the attributes of being polygon, three-sided, three-angled, and so on. It is not unreasonable to assume (at least according to Aristotle and many other traditional and modern logicians) that a substantive term like 'man' has essential properties as well. Every man must then have these properties. But what would be the essential properties of a negative term like 'not man'?[2] It is well-known that if we just extend syllogistic logic with negative terms, we have to add to our proof system a double negation rule ($T \equiv \overline{\overline{T}}$) and contraposition ($SaP \vdash \overline{P}a\overline{S}$). To be able to reason also with conjunctive terms does *not* mean that we have to assume all axioms of Boolean algebra, for Aristotle did not assume the existence of 'empty' and 'universal' terms, i.e., terms which on an extensional interpretation denote the empty set and the whole domain, respectively.

3 Leibniz' Semantic Calculus

Just as some other scientists in the seventeenth century, also Leibniz conceived of a *characteristica universalis*, an "algebra" capable of expressing all conceptual thought. This algebra would include rules for symbolic manipulation, what he called a *calculus ratiocinator*. His goal was to put reasoning on a firmer basis by reducing much of it to a matter of calculation that many could grasp. The task of the universal characteristic would not only be to express the formal structure of valid deductive reasoning, but also to express the *content* of the ideas being reasoned about. Thus, it should be a *semantic* calculus. The characteristic would build on an alphabet of human thought, a set of unanalyzable primitive meaningful concepts. Moreover, the characteristic should be *compositional*: any character representing a complex concept should correspond to the composition of the complex concepts from its simpler conceptual parts. In the late 1670s, Leibniz worked on a type of characteristics that satisfies both of these requirements. It was partly inspired by Descartes' dream of a 'universal mathematics' and Hobbes' idea that reasoning was *literally* like numerical calculation in arithmetic.

Leibniz had made several attempts to arithmetize the syllogism, i.e., to find arithmetic translations of the four propositional types of the square of opposition that would make all the valid assertoric moods into truths of arithmetic and all the invalid ones into arithmetic falsities. In one of the first trials, he uses prime numbers to symbolize elementary concepts. The reason why prime numbers are interesting is that for each number there is a unique way of expressing it as a product of prime

[2] For some adjectives (like 'tall' and 'heavy') it seems less unreasonable to propose that their negative counterparts have essential properties, but it is perhaps no accident that in natural language these negative counterparts are expressed positively by their antonyms (like 'short' and 'light').

numbers. Let us say that an elementary concept T is symbolized, or interpreted, by $f(T)$. By ignoring negative concepts, and by representing 'conjunctive' concepts in the form of products of elementary concepts, he so associated a numerical characteristic with each concept. Leibniz wrote the universal affirmative proposition SaP in the form $f(S) = f(P) \times y$, i.e. $\exists y(f(S) = f(P) \times y)$, or $f(P)$ divides $f(S)$ $(f(P)|f(S))$, and particular affirmative propositions (SiP) as $\exists x, y[f(S) \times x = f(P) \times y]$. Universally negative propositions (SeP) and particular negative propositions (SoP) are negations of SiP and SaP, respectively. A major problem of this system was that propositions of the form SiP are much too easily true (and propositions of the form SeP much too easily false). One way to improve this system tried by Leibniz was to say that SiP is true iff $f(S)$ and $f(P)$ have a greatest common divisor greater than 1, i.e. $gcd(f(S), f(P)) > 1$. Look, for instance, at the syllogistic argument $MeP, SaM \vdash SeP$. This reasoning is valid iff if $gcd(f(M), f(P)) = 1$ and $(f(M)|f(S))$, then $gcd(f(S), f(P)) = 1$. Because the latter sentence is a truth of arithmetic, the argument is valid. Unfortunately, this new method mispredicts for certain valid syllogisms. Consider Darii: $MaP, SiM \vdash SiP$, and assume that $f(M) = 2 \times 3, f(P) = 3$, and $f(S) = 2 \times 5$. Although $f(P)$ divides $f(M)$ and $f(S)$ and $f(M)$ have a common divisor larger than $1, f(S)$ and $f(P)$ have no such common divisor. Still, Darii is a valid syllogism.

As explained by Sotirov [26], the cause of the problem was Leibniz' confusion of intensional and extensional interpretations of terms. According to a set theoretic *extensional* interpretation of terms, terms are just interpreted as sets of individuals. A sentence of the form SaP is true iff all individuals in the extension of S are also individuals in the extension of P. According to the (dual) set theoretic *intensional* interpretation, terms stand for concepts, thought of as sets of attributes. A sentence of the form SaP is considered to be true iff each attribute associated with P is also associated with S. The extensional interpretation was favored by scholastic logicians like Ockham, but Leibniz clearly favored the intensional interpretation (and claims that Aristotle did so as well).[3]

Aristotle clearly assumes that essential predication is stronger than accidental predication, and medieval logicians very much assumed the same.[4] The problem

---

[3] The intensional view is also explicitly discussed in Wittgenstein's Tractatus, 5.1222: if p follows from q, then the sense of p is contained in the sense of q.

[4] According to Leibniz, Aristotle, in contrast to a nominalist like Locke, preferred the intensional interpretation:

> *Philalethes* (expressing Locke's view) [...] it appeared to me preferable to reverse the order of the premisses of syllogisms, and to say: *All A is B, all B is C, so all A is C*, rather than saying *All B is C, all A is B, so all A is C.* [...]
> *Theophilus* (expressing Leibniz's view) [...] Aristotle may have had a special reason for adopting [what is now] the common arrangement. For rather than saying 'A is B' he usually says 'B is in A' [...]. And with that way of stating it he achieves, through the accepted arrangements, the very connection which you insist upon. For instead of saying 'B is C, A is B, so A is C', Aristotle will express it thus: 'C is in B, B is in A, so C is in A'. For instance, instead of saying 'Rectangles are isogons (i.e. have equal angles), squares are rectangles, so squares are isogons', Aristotle will put the 'middle term' in the middle position without changing the order of the propositions, by stating each of them in a manner which reverses

of the above arithmetic analysis was that universal affirmative propositions were interpreted intensionally (*SaP* is true iff $f(P)$ divides $f(S)$, meaning that every attribute of *P* is also an attribute of *S*), while particular affirmative propositions were interpreted extensionally (*SiP* is true iff $f(S)$ and $f(P)$ have a common divisor, i.e., iff $f(S)$ and $f(P)$ have a non-empty intersection).[5] Having noticed the problem, it is easy to see that in principle there are two ways to solve it (cf. [26]): either to give both types of sentences an extensional, or both types of sentences an intensional interpretation. According to the first solution one says that *SaP* is true iff $f(S)$ divides $f(P)$, i.e. $(f(S)|f(P))$, meaning that every divisor of $f(S)$ is also a divisor of $f(P)$. The second solution is somewhat more complicated, because integers have a least number—1, but not a largest number. To still give an intensional interpretation to particular affirmative propositions, one can introduce an arbitrary integer n greater than 1 such that the interpretation of each term T is a proper divisor of n, i.e., $(f(T)|n)$. Now Sotirov proposes to interpret *SiP* as true iff the least common multiple of $f(S)$ and $f(P)$ is less than n. One can show that both of these solutions work: valid syllogisms are turned into truths of arithmetic and invalid ones into arithmetic falsities.[6] This proof is based on the fact that the operations of greatest common divisor, least common multiple, and division into n (that is, $\overline{x} = n/x$), can be shown to satisfy all the Boolean laws when their arguments range over all sets of prime factors of n, with union corresponding to least common multiple, intersection to greatest common divisor, and complement to division into n. This is a standard result, if n is square-free (i.e., n is not divisible by any square greater than 1).

Leibniz's own solution to the problems of the earlier system was different, but worked as well. His proposal was fully in the spirit of the intensional analysis. He interpreted each term T by a *pair* of numbers, $\langle f_1(T), f_2(T)\rangle$ such that $gcd(f_1(T), f_2(T)) = 1$ (the numbers $f_1(T)$ and $f_2(T)$ are called 'relatively prime'). Intuitively, T contains the set of basic attributes corresponding to the prime factors of $f_1(T)$, while T does (definitely) not contain the set of basic attributes corresponding to the basic factors of $f_2(T)$. The universal affirmative proposition of the form *SaP* is now considered as true iff $f_1(P)$ divides $f_1(S)$, i.e. $(f_1(P)|f_1(S))$, and $f_2(P)$ divides $f_2(S)$, i.e. $(f_2(P)|f_2(S))$. To illustrate, the sentence 'All men are rational' is true iff every attribute that definitely belongs to 'rationality' definitely belongs to

(Footnote 4 continued)
the order of terms, thus: 'Isogon is in rectangle, rectangle is in square, so isogon is in square'. This manner of statement deserves respect; for indeed the predicate is in the subject, or rather the idea of the predicate is included in the idea of the subject. [...] The common manner of statements concerns individuals, whereas Aristotle's refers rather to ideas or universals. [...]
Leibniz, New Essays on Human Understanding, Book 4, Chap. 17, Sect. 8)

[5] Having a non-empty intersection of the intensions of *S* and *T* is not enough for the sentence *SiP* to be true: although both gold and silver clearly share a property (e.g. being a metal) this doesn't mean that there is something both pure gold and pure silver.

[6] Glashoff [10] rightly complains that Sotirov's solution is not completely in the spirit of Leibniz's assumptions: Leibniz asumed that the builiding blocks (the prime numbers) can be an infinite set. This is impossible with Sotirov's solution.

'manhood' and every attribute that definitely doesn't belong to rationality definitely doesn't belong to manhood either. The particular affirmative proposition *SiP* is said to be true iff neither pair of non-corresponding numbers have a common divisor greater than 1, i.e., iff $gcd(f_1(S), f_2(P)) = 1$ and $gcd(f_2(S), f_1(P)) = 1$. Intuitively this means that the intensions of *S* and *P* are consistent with each other. Leibniz [13] showed that under this method all the laws of conversion and of the square of opposition are predicted to be valid. Łukasiewicz [20] showed that all the valid moods of the assertoric syllogism and the rule of *Reductio per impossible* are predicted to be valid as well.[7]

After the 1670s Leibniz hardly ever came back to his arithmetization of logic. It is not completely clear why. But there is room for speculation. The first speculation (see [3]) concerns the necessity to use (prime) numbers in the first place. As said above, Leibniz' goal was to find arithmetic translations of all terms such as to reduce validity to arithmetic truth. What he end up doing was not quite like that. To see this, let us see when an argument should be counted as valid. That is, when should we count $\phi_1, \ldots, \phi_n/\psi$ as a valid argument? Let ϕ be a sentence of the form *SaP*, *SiP*, *SeP*, or *SoP*, and say, for instance, that $f(SiP) = 1$ iff $gcd(f_1(S), f_2(P)) = 1$ and $gcd(f_2(S), f_1(P)) = 1$. Then, $\phi_1, \ldots, \phi_n \models \psi$ iff for all f: if $f(\phi_1) = 1$ and $\cdots$ and $f(\phi_n) = 1$, then $f(\psi) = 1$. Thus, it is not enough for a syllogism to be valid if there exists an arithmetic interpretation according to which the premises are true and the conclusion is true as well. In fact, Couturat [5] had reason to believe that Leibniz himself found an arithmetic interpretation of an invalid syllogism that corresponds to an arithmetic truth.[8] Leibniz' dilemma was then to either find the correct *unique* arithmetic interpretation for which arithmetic truth would always correspond to logical validity, or to give up on any uniqueness claim. The first horn of the dilemma means that even before we can start to calculate, we first have to work out the complete *characteristica universalis*. The second horn would involve being content with a much more modest *calculus ratiocinator*. Although he never

---

[7] Sommers [25] proposed an alternative numerical way to account for for syllogistic reasoning without making use of prime numbers, and more in the spirit of the medieval distribution theory. Unfortunately, Sommers' numerical method alone doesn't quite do the job. He needs an additional non-numerical rule: the requirement that for a syllogism to be valid, the number of particular conclusions must equal the number of particular premises. Friedman [8] improved on Sommers' method by getting rid of this additional rule. In fact, he showed that there are at least two purely numerical ways to account for syllogistic reasoning. According to the *additional* method one should replace *SaP* by $-S + P$, *SiP* by $+S + P$, *SeP* by $-S - P - 1$, and *SoP* by $+S - P - 1$. Let ϕ' be the result of the replacement of sentence ϕ. Then one can show that $\phi_1, \cdots, \phi_n \vdash \psi$ iff $\phi'_1 + \cdots + \phi'_n = \psi'$. According to the *multiplicational* method we replace *SaP* by $\frac{P}{S}$, *SiP* by $2SP$, *SeP* by $\frac{-1}{SP}$, and *SoP* by $\frac{-2S}{P}$. If we denote the result of the replacement in this way by ϕ", it follows that $\phi_1, \cdots, \phi_n \vdash \psi$ iff $\phi"_1 \times \cdots \times \phi"_n = \psi"$. Both methods validate all and only the valid syllogism, but the multiplicational method has an advantage because it allows for a natural representation of negative terms: $\overline{P}$ is represented as $\frac{1}{P}$. Not using prime numbers makes the calculations easier, but note that the resulting systems are anything but a characteristics universalis. In fact, the resulting systems cannot be thought of as semantic systems at all.

[8] Let M be assigned $\langle 10, 3\rangle$, S be $\langle 8, 11\rangle$, and P be $\langle 5, 1\rangle$. On this assignment, the syllogism *MaP*, *MoS*/*SoP* is wrongly predicted to valid.

seems to have given up hopes to find a fully universal characteristics, in later work he practically limited himself to the more mundane project. But for working out the second project, there is no essential reason to make use of (prime) numbers.

A second reason for why Leibniz might have given up on his arithmetization-project is that he was unable to extend it so as to include negative terms [9]. He realized, for instance, that in case T is a complex term represented by $\langle n, m\rangle$, $\overline{T}$ cannot be represented simply by $\langle m, n\rangle$: that the complex term 'white paper' does definitely not have the property of being black, doesn't entail that being something that is not a white paper means that it *must be* black. Leibniz tried other options for representing negative terms by pairs of numbers that are relatively prime (i.e. have no common divisor) in terms of his representation of positive terms, but failed. Glashoff [10] recently showed that this is no coincidence: he *could not* succeed if he wanted to satisfy the law of contraposition ($SaP \models \overline{P}a\overline{S}$). Glashoff [10] shows that Leibniz's artithmetical project *can* be saved, if we allow for 'richer' types of numbers.

Recall that what Leibniz did when he assigned numbers to terms was that he thereby interpreted these terms, i.e. he gave them a semantics. Nowedays we are more familiar with set theoretical models. Glashoff [11] recently provided a modern set theoretic intensional interpretation of Syllogistics following the spirit of Leibniz' final solution: interpreting terms by *pairs*. This solution is very interesting. Still, it would be helpful to see whether we can also give a 'modern' intensional interpretation more in line with Leibniz' earlier trials. Later in this paper I want to provide a natural set theoretic intensional interpretation of terms along these lines.[9] But before we do that, let us first see what is really required to give a semantics for syllogistic reasoning by looking at things algebraically.

4 Algebraic Semantics

Above, we have given an extensional and an intensional interpretation of terms. Following Leibniz, these interpretations were arithmetic in nature. We have seen, however, that interpretation was not essentially arithmetic at all. In fact, it doesn't matter much how terms are interpreted, as long as the interpretation is compositional. The extensional *arithmetic* interpretation discussed above corresponds closely with the standard *set theoretic* interpretation, according to which the extensional interpretation function E assigns to each primitive term T a non-empty subset of the set of objects D: $\emptyset \neq E_M(T) \subset D$. The sentence SaP is true iff $E(S) \cap E(P) = E(S)$ iff $E(S) \subseteq E(P)$, and SiP is true iff $E(S) \cap E(P) \neq \emptyset$. SoP and SeP are interpreted as the negations of SaP and SiP, respectively.

Of course, set theory comes with al the axioms of Boolean algebra, and this much structure is not at all required to model syllogistic reasoning. For the traditional fragment without conjunctive and negative terms, for instance, a partially ordered set $\langle U, \leq\rangle$, together with a relation @ which is reflexive and monotonic w.r.t. $\leq$ is already

[9] My proposal is thus closer to Sotirov's [26] approach.

enough (in fact, $\leq$ need only be a quasi-ordering, i.e., reflexive and transitive).[10] On the extensional interpretation, *SaP* is true iff $E(S) \leq E(P)$ (with $E(S) \in U$ being the extensional interpretation of *S*) and *SiP* is true iff $E(S)@E(P)$. For the intensional semantics, we demand that *SaP* is true iff $I(S) \geq I(P)$ and *SiP* is true iff $I(S)@I(P)$ (with $I(S)$ as the intensional interpretation of *S*). On the extensional interpretation we assume that $inf\{x, y\} \in U$ for all $x, y \in U$. On the intensional interpretation we assume, instead, that $sup\{x, y\} \in U$ for all $x, y \in U$. Alternatively, we can start for the extensional interpretation with a meet semi-lattice $\langle U, \bullet \rangle$, and determine $\langle U, \leq \rangle$ by defining $inf\{x, y\}$ as $inf\{x, y\} = x \bullet y$. Of course, we can also start with a join semi-lattice $\langle U, \circ \rangle$ and determine $\langle U, \leq \rangle$ by definition $sup\{x, y\} = x \circ y$. To interpret particular sentences, we can add a special element 0 to both types of semi-lattices. For the extensional interpretation the special element 0 is such that $\forall T \in Term$: $0 \bullet E(T) = 0$. Of course, *SaP* is true iff $E(S) \leq E(P)$ iff $E(S) \bullet E(P) = E(S)$ and *SiP* is true iff $E(S) \bullet E(P) \neq 0$ (the truth-conditions of *SoP* and *SeP* are determined as usual in terms of the truth-conditions of *SaP* and *SiP*). For the intensional interpretation, we start with the join semi-lattice. *SaP* is true iff $I(P) \leq I(S)$ iff $I(S) \circ I(P) = I(S)$ and we say that *SiP* is true iff $I(S) \circ I(P) \neq 0$, but now 0 should be thought of as the greatest element: $\forall T \in Term : 0 \circ I(T) = 0$. For simplicity we have assumed that we have only one 0-element. But this is not really required. Things would have worked as well if we had a *set* of minimal, or maximal, elements (called $\perp$) and we would have demanded that *SiP* is true on the extensional and intensional interpretation iff $E(S) \bullet E(P) \in \perp$ and $E(S) \circ E(P) \in \perp$, respectively.

Once we have '$\bullet$' and '$\circ$' as operations between terms, we can also account for conjunctive terms. Notice that on the intensional interpretation it is not '$\bullet$' but rather '$\circ$' that we use to interpret conjunctive terms. If we add term-negation to the language, we take $\langle U, \leq \rangle$ to be a distributive lattice and assume that for all $x, y \in U : \overline{\overline{x}} = x$ and $x \leq y \rightarrow \overline{y} \leq \overline{x}$. Observe that this is exactly a DeMorgan algebra if we don't have to assume that we have a *unique* minimal (or maximal) element. If we also want to interpret empty and universal terms, we assume that we have a whole Boolean algebra with unique minimal and maximal elements.

5 Lenzen's [18] Intensional Semantics

To give a set theoretic intensional semantics for syllogistics without negative terms, we have to start at least with a primitive set of attributes $\mathscr{A}$ and an interpretation function that assigns sets of attributes to terms. It is quite clear how to provide a semantics for sentences of the form *SaP*: $I(P) \subseteq I(S)$, but the problem is how to provide a semantics for particular sentences: *SiP*. The first idea that came to Leibniz's mind given in set-theoretic terms would be to say that *SiP* is true iff $I(S) \cap I(P) \neq \emptyset$. But this idea is clearly non-sensical: some bike is red, but there is nothing in the intension of 'red' that is also in the intension of 'bike', or so it seems. Or even more

[10] This is very well known, see, for instance [21, 26].

obviously, the sentence 'No gold is silver' is obviously true. According to the above suggestion this is true iff there is no attribute, or property, that gold and silver share. But there is obviously one: metal. What has to be assumed, rather, is the following idea: for 'Some bike is red' to be true, the intensions of 'red' and 'bike' should *not* be *incompatible*. On the other hand, for 'No silver is gold' to be true, the intensions of 'silver' and 'gold' should be *incompatible*. How should we model the intension of a term and of this notion of incompatibility?

Lenzen's [18] idea was to think of attributes as sets of individuals, and thus to think of the meaning of terms as *sets of sets* of (possible) individuals. He counted *SaP* as true iff $V_L(P)$ (the intensional Lenzen-interpretation) is a subset of $V_L(S) : V_L(P) \subseteq V_L(S)$. This is exactly as one would expect. For the interpretation of particular sentences of the form *SiP*, however, Lenzen makes use of the interpretation of the conjunctive term '*SP*': *SiP* is true iff $V_L(SP) \neq \wp(D)$, with D the set of all possible individuals.[11] The sentences *SoP* and *SeP* are interpreted as usual, meaning that *SeP* is true iff $V_L(SP) = \wp(D)$. But how does Lenzen interpret conjunctive terms? He interprets them as follows:

- $V_L(TT') = \{X \subseteq D : \bigcap V_L(T) \cap \bigcap V_L(T') \subseteq X\}$

Thus, for Lenzen [18], the interpretation of a conjunctive term is not the intersection of the interpretations of the two terms, but rather the intersection of their intersections. Recalling that according to Lenzen *SeP* is true iff $V_L(SP) = \wp(D)$, this means that *SeP* is true iff $\bigcap V_L(S) \cap \bigcap V_L(P) = \emptyset$.

Let us now consider a language which also has negative terms. How should we interpret this? It is straightforward to interpret such terms from an extensional point of view: $V_E(\overline{T}) = D - V_E(T)$, with D a set of individuals and $V_E(T)$ the elements of D that have property T. Things are more complicated when we look at things from an intensional perspective. We have seen that Leibniz was never able to give a satisfactory semantics for a language with both negative and conjunctive terms in terms of characteristic numbers. But Lenzen [18] provided an intensional semantics that is formally satisfactory.

The idea to intensionally interpret terms as sets of sets of individuals plays a major role. However, he demanded that not just any set of sets of individuals will do. If T is a term, Lenzen requires that the intensional Lenzen-interpretation of T, $I_L(T)$, is a *proper filter*: (i) $I_L(T) \neq \emptyset$, (ii) if $X \in I_L(T)$ and $X \subseteq Y$, then $Y \in I_L(T)$, and (iii) if $X \in I_L(T)$ and $Y \in I_L(T)$, then $(X \cap Y) \in I_L(T)$. We have seen already how he interprets conjunctive terms and simple sentences, but here is the full definition:

- $V_L(T) = I_L(T)$, if T is a primitive term.
- $V_L(TT') = \{X \subseteq D : \bigcap V_L(T) \cap \bigcap V_L(T') \subseteq X\}$.
- $V_L(\overline{T}) = \{X \subseteq D : \overline{\bigcap V_L(T)} \subseteq X\}$.

[11] If we think of the extensional counterpart, this means that 'some bike is red' is true not because there actually exists a red bike, but rather that it is possible that such a bike exists. And indeed, what Leibniz considers to be the extension of a term (a set of individuals scattered around all worlds) is very much what in possible worlds semantics is its intension (cf. [13] and [12, p. 49]).

- $V_L(SaP) = 1$ iff $V_L(P) \subseteq V_L(S)$
  $V_M(SiP) = 1$ iff $V_L(SP) \neq \wp(D)$
  SoP and *SeP* are interpreted as usual.

Entailment is defined as usual. Lenzen [18] shows that this interpretation is in fact equivalent to the set theoretic extensional interpretation in the sense that it validates exactly the same inferences. This is very pleasing, just as the fact that for all terms S and T their intension and extension behave as duals: $V_L(S) \supseteq V_L(T)$ iff $V_E(S) \subseteq V_E(T)$.

Still, I believe that there is reason to be unsatisfied with Lenzen's intensional semantics. The reason is that many authors (e.g. [3, 9, 11]) naturally assume that for a semantics to be called intensional it should not be the case that we have to make reference to individuals. But exactly this reference to individuals is crucial for Lenzen's interpretation of conjunctive and negative terms: he crucially thinks of attributes as sets of individuals. What we would like to have instead is to assume with Leibniz that basic attributes are just primitives. But how should we proceed?

6 Towards a Truly Intensional Semantics

6.1 An Intensional Semantics for Simple Syllogistics

We have seen in the previous section that to account for sentences of the form *SiP* and *SeP* we have to make sense of intensions of terms being compatible or incompatible with each other. Lenzen [18] idea to do this was to make use, in the end, of (possible) individuals. But we have argued above that this is unsatisfying when one wants to provide a 'truly' intensional semantics. In a truly intensional semantics, one rather starts with a set of attributes, $\mathscr{A}$, as being primitive, and not defined in terms of sets of individuals. But once one does so, one also has to assume that the notion of (in)compatibility is primitive as well.

The fact that we have to assume such a primitive notion of (in)compatibility already suggests why Leibniz had a hard time to come up with a satisfying characteristics for even simple syllogistic logic. Just like Wittgenstein when he was writing his Tractatus, also Leibniz thought of his simple terms, or attributes, as being *logical independent* of each other, i.e., their being mutually compatible with all other simples (cf. [12, p. 54]): only if the simples are logically independent of each other is it possible to construct a language where inference and equivalence can be checked 'from the surface'. To check validity we don't have to know what the interpretation of the different terms is. But if all terms are interpreted by sets of these simple attributes that are all mutually compatible with each other, a sentence like 'No gold is silver' can never be true.

In our first interpretation, we will intensionally interpret each term as a set of attributes. However, these attributes need not all be simple, i.e., we don't demand that $I_M(T) \subseteq \mathscr{A}$, if I is the interpretation function. Rather, we assume a primitive operator '$\circ$', such that $\langle \mathscr{A}^*, \circ \rangle$ is a semi-lattice, i.e., the elements of $\mathscr{A}^*$ are closed under '$\circ$', and are generated by the set of primitive features $\mathscr{A}$. We will assume that each term denotes a subset of $\mathscr{A}^*$ closed under '$\circ$'. Furthermore, we assume that we have a primitive set of inconsistent (or impossible) attributes, called $\perp$. To accounts for existential import we demands that $\forall x \in I_M(T) : x \notin \perp$ for all terms T.[12]

We interpret the language with respect to a model $M = \langle \mathscr{A}, \circ, I, \perp \rangle$. As said above, $\mathscr{A}$ is the set of primitive 'simple' attributes, and '$\circ$' is an operator that is commutative, associative, and idempotent. The set of attributes $\mathscr{A}^*$ is generated by, and thus a superset of, $\mathscr{A}$. I is an interpretation function which assigns to each primitive term T a subset of $\mathscr{A}^*$ (i) closed under $\circ$ and (ii) no element of $I_M(T)$ is an element of $\perp$. We say that $I_M(SaP) = 1$ iff $I_M(S) \supseteq I_M(P)$ and $I_M(SiP) = 1$ iff $\neg\exists x \in I_M(S) : y \in I_M(P) : x \circ y \in \perp$. Thus, *SiP* is true iff *S* and *P* do not contain mutually incompatible attributes. *SoP* and *SeP* are interpreted as true iff *SaP* and *SiP*, respectively, are not.

We say, as usual, that $\Gamma \models \psi$ iff $\forall M$: if $\forall \phi \in \Gamma : V_M(\phi) = 1$, then $V_M(\psi) = 1$. This semantics validates all and only all arguments in Aristotelian syllogistic style if and only if they are counted as valid on the standard extensional semantics where *S* and *P* denote non-empty sets of individuals, and *SaP* and *SiP* are true iff $E(S) \subseteq E(P)$ and $E(S) \cap E(P) \neq \emptyset$, respectively. The proof of this makes use of the fact that this standard fragment can be axiomatised by the validity of (i) *SaS*, (ii) *SiS*, and the syllogisms (iii) Barbara and (iv) Datisi (cf. [20]). But these validities immediately follow because (a) the '$\subseteq$'-relation between sets of attributes gives rise to a partial order and (b) the relation of 'compatible union' between the interpretation of primitive terms is reflexive and downward monotonic with respect to '$\subseteq$'.

6.2 An Intensional Semantics for Syllogistics with Complex Terms

How can we provide an intensional interpretation of syllogistics with conjunctive and negative terms without crucially referring to individuals? One way to go would be to use a similar trick as in the previous section, but now don't start with the semi-order $\langle \mathscr{A}^*, \circ \rangle$, but rather with a distributive lattice $\langle \mathscr{A}^*, \circ, \star \rangle$, which also has a complementation operaton, which satisfies (i) double negation, $\overline{\overline{x}} = x$, and the DeMorgan laws: $\overline{x \circ y} = \overline{x} \star \overline{y}$ and $\overline{x \star y} = \overline{x} \circ \overline{y}$. If we define $x \leq y$ as usual: $x \leq y$ iff $x \circ y = y$ iff $x \star y = x$, it is easy to prove that it follows that $x \leq y$ iff $\overline{y} \leq \overline{x}$.[13] This can then be used to account for interpreting conjunctive and negative terms.

Such an account would not really fulfill our goal to find a purely set-theoretic semantics, however, because we still would have used an algebraic semantics. What if we just want to work with attributes and sets thereof? The straightforward way to go is to *lift* the interpretations of our earlier intensional model: terms should not

---

[12] The idea behind these constraints is similar to [18] idea to assume that each term intensionally denotes a proper filter.

[13] Proof: $x \leq y$ iff $x \star y = x$ iff $\overline{x \star y} = \overline{x}$ iff $\overline{x} \circ \overline{y} = \overline{x}$ iff $\overline{y} \leq \overline{x}$.

be interpreted as *sets* of attributes, but rather as *sets of sets* of (simple) attributes. In that case we don't have to assume that the attributes themselves are closed under some operators like '$\circ$' or '$\star$'. Instead, we can now assume that we start with a model $M = \langle \mathscr{A}, I, \perp \rangle$, where $\mathscr{A}$ is a set of simple attributes, I an interpretation function which assigns to each primitive term T a consistent subset of $\mathscr{A}$, and where $\perp$ is now a primitive symmetric and irreflexive relation between elements of $\mathscr{A}$. We now make the assumption that for every $x \in \mathscr{A} : \exists! y \in \mathscr{A} : x \perp y$. We define $\mathscr{A}^*$ as the set of all consistent subsets of $\mathscr{A}$: $\{X \subseteq \mathscr{A} : \neg\exists x, y \in X : x \perp y\}$. Let I an interpretation function which assigns to each primitive term T an element of $\mathscr{A}^*$, i.e., a set of mutually consistent attributes. In general, however, terms will be interpreted as *sets* of sets of attributes. The intensional interpretation of terms T, $V_M(T)$, is recursively defined as follows (inspired by van Fraassen [7]):

- $V_M(T) = \{I_M(T)\}$, if T is a primitive term.
- $V_M(TT') = \{X \cup Y : X \in V_M(T)\ \&\ Y \in V_M(T')\}$
- $V_M(\overline{T}) = \bigwedge\{\overline{X} : X \in V_M(T)\}$, with $\overline{X} = \{\{\overline{x}\} : x \in X\}$,
  and $\mathscr{X} \wedge \mathscr{Y} = \{X \cup Y : X \in \mathscr{X}\ \&\ Y \in \mathscr{Y}\}$

- $\mathscr{F} \sqsubseteq \mathscr{G}$ iff $\forall X \in \mathscr{F} : \exists Y \in \mathscr{G} : Y \subseteq X$
- $\mathscr{F} @ \mathscr{G}$ iff $\exists X \in \mathscr{F} : \exists Y \in \mathscr{G} : X \cup Y \in \mathscr{A}^*$

- $V_M(SaP) = 1$ iff $V_M(S) \sqsubseteq V_M(P)$
  $V_M(SiP) = 1$ iff $V_M(S) @ V_M(P)$

- $\Gamma \models \psi$ iff $\forall M$: if $\forall \phi \in \Gamma : V_M(\phi) = 1$, then $V_M(\psi) = 1$.

Notice that the analysis implements a 'truly' intensional semantics because (i) it doesn't make use of individuals, and (ii) because of the (modified) 'predicate is in the subject'-analysis of universal statements. It is easy to prove that '$\sqsubseteq$' is both reflexive and transitive. Still, it does not give rise to a partial order, because it is not antisymmetric. For instance: $\{\{t\}, \{t, s\}\} \sqsubseteq \{\{t\}\}$ and $\{\{t\}\} \sqsubseteq \{\{t\}\{t, s\}\}$, but obviously $\{\{t\}\} \neq \{\{t\}, \{t, s\}\}$. As example discussed by Leibniz [15] we can mention the terms *triangle* and *trilateral*. We have seen in Sect. 4 that the fact that '$\sqsubseteq$' is both reflexive and transitive is enough to account for syllogistic logic without negative terms, if the relation '@' is both reflexive and monotone w.r.t. '$\sqsubseteq$'. Is that the case? It is easy to see that it is not, because '@' is not reflexive. As we will see later, this has important consequences.

But let us first try to understand the interpretation of terms a bit better. Let us assume that we don't have negative terms. In that case, the only complex terms that exist are 'conjunctive' terms. Because primitive terms always denote singleton sets, it follows that also all conjunctive terms denote singleton sets, and thus (without term negation) *all* terms denote singleton sets. In that case, our lifted interpretation is equivalent to our earlier intensional interpretation: If we would forget in that earlier interpretation 'closure under '$\circ$'', what we earlier had as $V_M(T)$ would now be $\bigcap V_M(T)$. It follows that *SaP* is true iff $\bigcap V_M(P) \subseteq \bigcap V_M(S)$, and *SiP* is true iff $\bigcap V_M(P) \cup \bigcap V_M(S) \in \mathscr{A}^*$. Thus, our new system is a conservative extension of our earlier system, and indeed, it is easy to prove that this semantics validates all and only all valid syllogisms without negative terms. But what about the extension with negative terms? What about double negation and contraposition?

To illustrate the treatment of negative terms, assume that $A = \{x, y, z, v\}$, and $x \perp z$ and $y \perp v$. We will denote z and v, respectively, by $\overline{x}$ and $\overline{y}$, because we assume that $\forall x \in \mathscr{A} : \exists! y \in \mathscr{A} : x \perp y$. In that case, the set of compatible sets of attributes is $\{\{x\}, \{y\}, \{\overline{x}\}, \{\overline{y}\}, \{x, y\}, \{x, \overline{y}\}, \{\overline{x}, y\}, \{\overline{x}, \overline{y}\}\}$. A term does not denote a consistent set of attributes, but rather a *set* of consistent sets of attributes. The singleton sets $\{\{x\}\}, \{\{y\}\}$ can be thought of as (the denotations of) simple properties. Call them X and Y, respectively. The compatible complex properties can then be denoted by $XY, X\overline{Y}, \overline{X}Y$, and $\overline{X}\overline{Y}$. Disjunctive properties typically denote non-singleton sets. The disjunctive property consisting of X and Y, for instance, denotes $\{\{x\}, \{y\}\} = V_M(X) \cup V_M(Y)$. Whether negative predicates denote singleton sets or not depends on how many incompatibles they have. Because x, for instance, is only incompatible with $\overline{x}$, $\overline{X}$ denotes the singleton sets $\{\{\overline{x}\}\}$. It is different with conjunctive properties like XY: $V_M(\overline{XY}) = \{\{\overline{x}\}, \{\overline{y}\}\} = V_M(\overline{X}) \cup V_M(\overline{Y})$. Notice that from these we can go back to the original via double negation: $V_M(\overline{\overline{X}}) = \{\{x\}\} = V_M(X)$ and $V_M(\overline{\overline{XY}}) = \{\{x, y\}\} = V_M(XY)$. As it turns out, this holds in general: for all terms T: $V_M(\overline{\overline{T}}) = V_M(T)$.

On the analysis so far, although a sentence like 'All square circles are circles' comes out true, the sentence 'All square circles are green' need not. In fact, in what we have now, much less follows than it does on an extensional semantics. This is so, because we make distinctions that an extensional semantics cannot make. In particular, we make a distinction between two contradictory terms like $P\overline{P}$ and $Q\overline{Q}$, and thus also to their 'tautological' disjunctive negations: $V_M(\overline{P\overline{P}}) \neq V_M(\overline{Q\overline{Q}})$. Intuitively, the reason is that $V_M(\overline{T}) \neq \wp(A) - V_M(T)$. In our toy model above $\{x, \overline{x}\} \neq \{y, \overline{y}\}$. This allows us to make a distinction between different inconsistent concepts, like *square circle* and *triangular circle*. The one can denote $\{\{x, \overline{x}\}\}$ while the other denotes $\{\{y, \overline{y}\}\}$. Similarly, $\{\{x\}, \{\overline{x}\}\} \neq \{\{y\}, \{\overline{y}\}\}$, meaning that two tautological concepts need not have the same meaning. What this illustrates is that our intensional analysis is more fine-grained than standard extensional semantics.[14] But allowing for contradictory concepts has two important consequences: one for our notion of truth, and one for our notion of consequence. First, sentences like SiS are not always true for all terms S, i.e., not if S is a contradictory term. This shows that the relation '@' is not reflexive. But this has as a consequence that syllogistic inferences that rely on existential import are by our semantics not predicted to be valid. Consider, for instance, Darapti, a syllogism of the third figure with the form '$MaP, MaS/SiP$'. Consider the example 'All square circles are square, all square circles are circles, thus some circles are square'. On our semantics given so far, both premisses are true, but the conclusion is not.

To overcome the problem I propose that we just limit ourselves in reasoning to non-contradictory terms. Limited to such terms, the relation '@' becomes

[14] In fact, we end up with something very close to the syllogistic counterpart of a semantics of Anderson and Belnap's [1] notion of tautological (or relevant) entailment. Indeed, I have based the semantics on some ideas of van Fraassen [6], which is used to gives a semantics for this logic.

immediately reflexive, and it is easy to see that it then also behaves monotonic w.r.t. '$\sqsubseteq$'. Thus, by limiting ourselves to non-contradictory terms, we have proved that our semantics validates all and only all valid syllogisms in the classic formulation. But what about syllogisms which contain negative terms? Well, in that case double negation has to be valid, just as contraposition. We have mentioned already that double negation holds. What about contraposition? Fortunately, one can prove (following basically appendix 1 of [7])[15] that for all terms T_1 and T_2: if $V_M(T_1) \sqsubseteq V_M(T_2)$, then $V_M(\overline{T_2}) \sqsubseteq V_M(\overline{T_1})$. Thus, we have reached our goal.

Notice that even if we limit ourselves to non-contradictory terms, our semantics is still not Boolean. It does not even satisfy all of the DeMorgan laws. How can we go from our semantics to a richer *DeMorgan* or even *Boolean* semantics? As it turns out, we can follow van Fraassen [7] and assign to each term a somewhat different interpretation, the *closure* of $V_M(T)$ under the superset relation: $V^c_M(T) = \{Y \in \mathscr{A}^* \mid \exists X \in V_M(T) : X \subseteq Y\}$. It is clear that the closure of $\{\{t\}\}$ and $\{\{t\}, \{t, s\}\}$ is the same. Let $\mathscr{Z}$ be a set of sets of attributes. We say that $\mathscr{Z}$ is a closed property iff $\mathscr{Z} = \mathscr{Z}^c$, where $\mathscr{Z}^c$ is the closure of Z. van Fraassen [7] shows that intersections, unions, and complementations of such closed properties are closed again, and in fact form a DeMorgan algebra. Now we can reformulate the truth conditions of our categorical sentences in terms of our 'closured' interpretations:

- $V_M(SaP) = 1$ iff $V^c_M(S) \subseteq V^c_M(P)$
  $V_M(SiP) = 1$ iff $V^c_M(S) \cap V^c_M(P) \neq \emptyset$

To go fully Boolean, what we need, obviously, is *one* contradiction and *one* tautology. To establish this, we can limit ourselves to the *maximal consistent elements* of the closures of the meanings of our terms as follows. First, we can define the set of its maximal elements of the *closure* of $V_M(T)$ as follows:

- $Max(V^c_M(T)) = \{X \in V^c_M(T) : \neg\exists Y \in V^c_M(T) : X \subset Y\}$.

Next, we define the truth conditions of sentences in terms of maximal elements:

- $V_M(SaP) = 1$ iff $Max(V^c_M(S)) \subseteq Max(V^c_M(P))$
  $V_M(SiP) = 1$ iff $Max(V^c_M(S)) \cap Max(V^cM(P)) \neq \emptyset$.

Observe that although $V_M(P\overline{P}) \neq V_M(Q\overline{Q})$ and $V_M(\overline{P\overline{P}}) \neq V_M(\overline{Q\overline{Q}})$ and also $V^*_M(P\overline{P}) \neq V^*_M(Q\overline{Q})$ and $V^*_M(\overline{P\overline{P}}) \neq V^*_M(\overline{Q\overline{Q}})$, still it holds that $Max(V^*_M(P\overline{P})) = Max(V^*_M(Q\overline{Q})) = \emptyset$ and $Max(V^*_M(\overline{P\overline{P}})) = Max(V^*_M(\overline{Q\overline{Q}}))$ = the set of all maximally consistent sets of attributes. Thus, there is only one contradiction and one tautology.

[15] See the same paper for a proof why double negation holds in our semantics.

7 Conclusion and Outlook

In this paper I have given an intensional semantics of syllogistic logic: a semantics for syllogistics logic which does not mention individuals. It is well established that the intensional view is in some sense the 'inverse' of the extensional one. Still, two challenges had to be met: (i) how to account for *i*- and *e*-sentences like 'Some/No philosophers are rich', and (ii) how to account for the combination of negative and conjunctive terms. To account for (i) we have taken a notion of *(in)compatibility* to be primitive. To account for (ii) we have proposed to *lift* the intensional interpretation of a term from a set of attributes to a set of sets of attributes. The latter idea was mostly based on a (loose) analogy with van Fraassen [6] analysis of the tautological entailment based on facts.

Syllogistic reasoning can be extended to relations (see e.g. [23–25]) and it is an interesting question whether also this extension can be given an intensional semantics. It is obvious that we need more than just take features to be basic. To account for this extension remains an interesting challenge for the future.

References

1. Anderson, A., & Belnap, N. (1975). *Entailment* (Vol. I). Princeton: Princeton University Press.
2. Aristotle, A. L. (1942). *Prior analytics*. New York: Oxford University Press.
3. Bassler, B. (1998). Leibniz on intension, extension, and the representation of syllogistic inference. *Synthese*, *116*, 117–139.
4. Bird, O. (1964). *Syllogistic and its extensions*.New Jersey: Prentice-Hall Inc.: Fundamentals of Logic Series.
5. Couturas, L. (1901). *La Logique de Leibniz*. Paris: Felix Alcan.
6. van Fraassen, B. (1969). Facts and tautological entailments. *Journal of Philosophy*, *66*, 477–487.
7. van Fraassen, B. (1973). Extension, intension, and comprehension. In M. Munitz (Ed.), *Logic and ontology* (pp. 101–131). New York: New York University Press.
8. Friedman, W. (1980). 'Calculemus', Notre Dame. *Journal of Formal Logic*, *21*, 166–174.
9. Glashoff, K. (2002). On Leibniz' characteristic numbers. *Studia Leibnitiana*, *34*, 161.
10. Glashoff, K. (ms), 'On negation in Leibniz' system of characteristic numbers', manuscript.
11. Glashoff, K. (2010). An intensional Leibniz semantics for Aristotelian logic. *The Review of Symbolic Logic*, *3*, 262–272.
12. Ishiguro, H. (1972). *Leibniz' philosophy of logic and language*. London: Duckworth.
13. Leibniz, G. (1966a). Rules from which a decision can be made, by means of numbers, about the validity of inferences and about the forms and moods of categorical syllogisms. In G. H. R. Parkinson (Ed.), *Leibniz: Logical papers* (pp. 25–32) Oxford: Clarendon Press.
14. Leibniz, G. (1966b). A paper on 'some logical difficulties'. In G. H. R. Parkinson (Ed.), *Leibniz: Logical papers* (pp. 115–121) Oxford: Clarendon Press.
15. Leibniz, G. (1966c). Of the mathematical determination of syllogistic forms. In G. H. R Parkinson (Ed.) *Leibniz: Logical papers* (pp. 105–111) Oxford: Clarendon Press.
16. Leibniz, G. (1973). The nature of truth. In G. H. R. Parkinson (Ed.) *Leibniz: Philosophical writings* (pp. 93–95).
17. Leibniz, G. (1996). New essays on human understanding. In: P. Remnant & J. Bennet (Eds.), *Cambridge texts in the history of philosophy*. Cambridge: Cambridge University Press.

18. Lenzen, W. (1983). Zur extensionalen und “intensionalen” interpretationen der Leibnizschen logic. *Studia Leibnitiana, 15*, 129–148.
19. Lenzen, W. (1990). *Das System der Leibniz'schen Logik*. Berlin: De Gruyter.
20. Łukasiewicz, J. (1951). *Aristotle's syllogistic from the standpoint of modern formal logic*. Oxford: Clarendon Press.
21. Martin, J. N. (1997). Aristotle's natural deduction reconsidered. *History and Philosophy of Logic, 18*, 1–15.
22. Parkinson, G. H. R. (1966). *Leibniz: Logic papers*. Oxford: Clarendon Press.
23. Pratt-Hartmann, I., & Moss, L. S. (2009). Logics for the relational syllogistic. *Review of Symbolic Logic, 2*, 647–683.
24. van Rooji, R. (2012). The propositional and relational syllogistic. *Logique et Analyse, 55*, 85–108.
25. Sommers, F. (1970). The calculus of terms. *Mind, 79*, 1–39.
26. Sotirov, J. (1999). Arithmetization of Syllogistic a la Leibniz. *Journal of Applied Non-classical Logics, 9*, 387–405.
27. Wittgenstein, L. (1933). *Tractatus Logico-Philosophicus*. London, New York: Kegan Paul, Trench, Trubner & Co.

Inter-Model Connectives and Substructural Logics

Igor Sedlár

Abstract The paper provides an alternative interpretation of 'pair points', discussed in [3]. Pair points are seen as points viewed from two different 'perspectives' and the latter are explicated in terms of two independent valuations. The interpretation is developed into a semantics using pairs of Kripke models ('pair models'). It is demonstrated that, if certain conditions are fulfilled, pair models are validity-preserving copies of positive substructural models. This yields a general soundness and completeness result for a variety of (positive) substructural logics with respect to multimodal Kripke frames with binary accessibility relations. In addition, an epistemic interpretation of pair models is provided.

Keywords Pair models · Substrucutral logics · Ternary semantics

1 Introduction

The recent paper [3] contains an interesting suggestion concerning the ternary semantics of substructural logics. It is suggested that the ternary relation may be seen as a binary relation between points and ordered pairs of points—*pair points*.[1] This paper develops the suggestion to a fully fledged semantics for (positive) substructural logics. The semantics dispenses with the ternary relation and uses a family of binary relations instead.[2] The paper is organised as follows. Section 2 explains the idea of pair points and explicates pair points by invoking a pair of *valuations*. Section 3 expands on the idea in a more formal manner. First, 'pair models' and 'pair

I. Sedlár (✉)
Comenius University in Bratislava, Bratislava, Slovakia
e-mail: sedlar@fphil.uniba.sk

[1] The suggestion appears also in [2, Chap. 2]. It builds upon a remark by [9].

[2] Hence, our investigations are similar in spirit to [5, 7]. See also [12], which has been shown to contain a flaw by [6], and the corrected version [13].

R. Ciuni et al. (eds.), *Recent Trends in Philosophical Logic*,
Trends in Logic 41, DOI: 10.1007/978-3-319-06080-4_14,

frames' are introduced. It is explained that these correspond to pairs of binary Kripke models and frames, respectively. Second, it is shown that, if certain conditions are fulfilled, pair models are validity-preserving copies of positive substructural models. These conditions constitute the notion of a pair model (frame) *representation* of a substructural model (frame). A general soundness and completeness result is established: numerous (positive) substructural logics are sound and complete with respect to specific classes of pair frames. Moreover, an independent characterisation of pair representations is provided. Third, it is explained that the result opens a new perspective on substructural connectives, in that the respective truth conditions refer to various models 'inside' pair models. Hence, substructural connectives have an 'inter-model' nature. Section 4 establishes an 'independent' completeness result for some very weak substructural logics. It is argued that if a logic is characterised by the class of all substructural frames, then it is characterised by the class of 'substructural pair models'. It is explained that the latter class is distinct from the class of pair model representations. Section 5 offers a 'philosophical' story behind our semantics. The story uses epistemic notions of communication and inference. Section 6 concludes the paper and points out the most important open problems.

2 Pair Points

We begin by explaining the idea of pair points in more detail. The motivation for considering pair points in [3] as well as the related technical details are outlined in Sect. 2.1. Afterwards, the original interpretation of pair points is explained and a new one is offered in Sect. 2.2. The new interpretation is the background of the technical work of Sect. 3.

2.1 Pair Points and Counterexamples

Beall et al. [3] observe that the substructural conditional, defined in terms of the ternary relation, seemingly bucks the 'no counterexample' interpretation. According to the interpretation, $A \to B$ is true at a point x iff there is no relevant counterexample y such that A is true at y whilst B is false at y. Quite so, since we can have $x \nVdash A \to B$ where there are *distinct* points y, z such that $Rxyz$, $y \Vdash A$, and $z \nVdash B$. In general, no *single* counterexample point that makes A true and B false is required to make a conditional $A \to B$ false at a given point.

Nonetheless, a simple way out is offered: the notion of a point is extended to include ordered pairs of 'old' points as well. More precisely, *pair points* $\langle xy \rangle$ are introduced, where x, y are 'old' points. In addition, R is rephrased as a binary relation between points and pair points: $Rxyz$ becomes $Rx\langle yz \rangle$. Consequently, pair points may serve as counterexamples for false conditionals.

However, two points need clarification. First, the truth condition of $A \rightarrow B$ should be explicitly restated in terms of pair points. This, in turn, requires a clear notion of truth and falsity at pair points. The authors of Beall et al. [3] proceed as follows. Truth $\models_T$ and falsity $\models_F$ at pair points are defined in terms of the original forcing relation $\Vdash$:

- $\langle xy \rangle \models_T A$ iff $x \Vdash A$
- $\langle xy \rangle \models_F A$ iff $y \nVdash A$

Hence, $\langle xy \rangle \models_T A$ is in general consistent with $\langle xy \rangle \models_F A$. However, this is not the case when one considers *half points* $\langle xx \rangle$, which can be identified with the 'old' points of the model. (A pair point $\langle xy \rangle$ where $x \neq y$ is called a *duo point*.)

The rephrased truth condition of $A \rightarrow B$ runs as follows:

- $x \models_T A \rightarrow B$ iff there is no $\langle yz \rangle$ such that $Rx\langle yz \rangle$ and $\langle yz \rangle \models_T A$, but $\langle yz \rangle \models_F B$
- $x \models_F A \rightarrow B$ iff $x \not\models_T A \rightarrow B$ (x may be seen as $\langle xx \rangle$)

It is plain that the conditional is now susceptible to the 'no counterexample' interpretation. If $A \rightarrow B$ is false at x, then there is a counterexample $\langle yz \rangle$, such that $Rx\langle yz \rangle$, $\langle yz \rangle \models_T A$, but $\langle yz \rangle \models_F B$.

2.2 Pair Points and Perspectives

Pair points yield an interesting reading of the ternary semantics and they bring it closer to the binary modal semantics. However, pair points need to be given an 'independent interpretation'. Otherwise, they shan't be anything more than a technical trick.

The original interpretation of pair points builds on the theory of situated inference, see [8]. Half points (or, equivalently, the 'old' points) are seen as situations in the style of [1] and duo points as *information links*. Examples of information links include 'laws of nature, conventions, and any information that gives us a license to make inferences' [3, p. 602].

But other interpretations are conceivable as well. Let us begin by considering the following example. Let T denote the statement that a given person a is tall. Now the two statements T and $\neg T$ may be seen as describing two different states of affairs (a as an adult and a as a child, for example). But it is also possible to see these statements as describing the *same* state of affairs from *two distinct points of view*. For example, a might be tall when compared with the rest of her family, but not tall when compared with the local basketball team.

The same observation applies to pair points. One approach is to see pair points as pairs of two (possibly distinct) situations. But there is also a quite different angle: the pair point $\langle xy \rangle$ may be seen as a *single* situation viewed from *two distinct perspectives*. For example, x, y may correspond to belief sets of two distinct agents in the situation $\langle xy \rangle$ (their different 'opinions' about the situation).

There is a simple way to model the difference between the above approaches. Let us, for the time being, consider only points and valuations without any reference

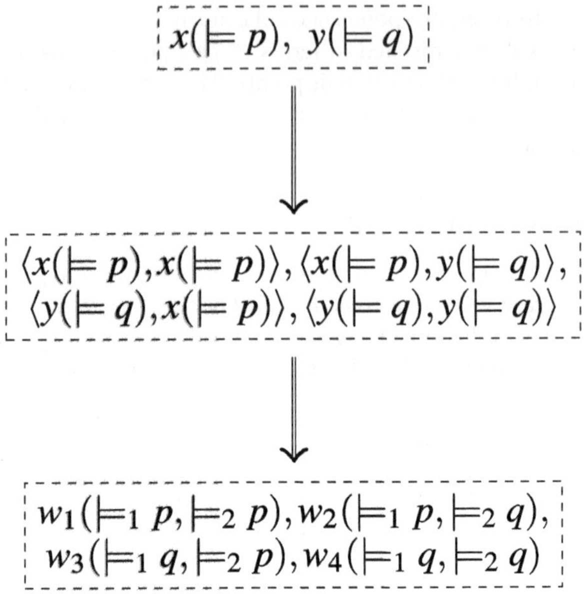

Fig. 1 Pair points and their simulation by means of two valuations

to accessibility relations. Take a pair (P, V), where P is a set of points and V is a valuation on P. The *pair point model* built on (P, V) is the pair (P^2, V). If we decided to extend (P, V) by a ternary relation, the relation could be simulated as a binary relation on P^2 in the pair point model.

But the pair point model might be simulated by a structure that does not include P^2. The key is to replace the somewhat vague 'perspectives' by *valuations*. The *pair representation* of (P^2, V) is a *triple* (P', V_1, V_2), such that there is a bijection σ from P' to P^2 and V_1, V_2 are valuations. Now $w \in P'$ might be seen as a *representation* of $\sigma(w)$ iff the following holds: $w_i \in V(p)$ iff $w \in V_i(p)$ for $i \in \{1, 2\}$, where w_i is the i-th member of $\sigma(w)$. Put differently, the 'binary nature' of $\langle xy \rangle \in P^2$ might be simulated by using two valuations V_1, V_2 'operating on' points that are not explicit pairs. For a simple example, see Fig. 1.

Now, obviously, we have an interesting twist in the story: pair representations might be seen as *pairs* of models of the original form: (P', V_1) and (P', V_2). Thus, pair point models correspond to pairs of 'simple models'. The following section develops this observation and applies it to substructural models.

3 Pair Representations

This section develops the interpretation of pair points, outlined in Sect. 2. Pair models and pair frames are defined and the truth conditions of substructural formulas are adapted to pair models. Pair model (pair frame) representations of substructural models (frames) are introduced in Sect. 3.1. Afterwards, a direct characterisation of pair representations is provided (Sect. 3.2). A general soundness and completeness result is then established: numerous (positive) substructural logics are sound and complete with respect to multimodal Kripke frames with binary accessibility relations (Sect. 3.3). It is explained that the result opens a new perspective on substructural connectives (Sect. 3.4).

3.1 The Basic Definitions

We shall be working with a positive substructural language $\mathscr{L}^+$, built upon a countable set of propositional variables Φ. Formulas of the language are built upon Φ by means of applying the binary operators '$\wedge$' (extensional conjunction), '$\vee$' (extensional disjunction), '$\rightarrow$' (implication), '$\circ$' (intensional conjunction, fusion) and '$\leftarrow$' (converse implication).

Validity shall be defined for *consecutions*, i.e. expressions of the form

$$X \vdash A$$

(read 'A is a consequence of X') where A is a formula and X is a *structure* built from formulas by means of applying the binary operations ',' (comma) and ';' (semicolon). For more details, see [10].

Definition 1 [10] A *(positive) substructural frame* is a triple

$$\mathfrak{F} = (P, \sqsubseteq, R)$$

where P is a non-empty set (of 'points'), $\sqsubseteq$ is a partial order on P and $R \subseteq P^3$.

A *(positive) substructural model* built on a frame $\mathfrak{F}$ is a couple

$$\mathfrak{M} = (\mathfrak{F}, \Vdash)$$

where $\Vdash$ is a forcing relation between points and members of Φ such that $x \sqsubseteq y$ and $x \Vdash p$ imply $y \Vdash p$ for all $p \in \Phi$.

The forcing relation can be extended to every formula and structure in a familiar way (see [10]). A consecution $X \vdash A$ is *valid in* $\mathfrak{M}$ iff $x \Vdash X$ implies $x \Vdash A$ for all $x \in \mathfrak{M}$, i.e. all x in P, where P belongs to $\mathfrak{M}$ (notation: $X \vdash_{\mathfrak{M}} A$). $X \vdash A$ is *valid in a frame* $\mathfrak{F}$ iff it is valid in every $\mathfrak{M}$ built on $\mathfrak{F}$ (notation: $X \vdash_{\mathfrak{F}} A$). If $\mathfrak{C}$ is a class of substructural frames, then $X \vdash A$ is *valid in* $\mathfrak{C}$ iff it is valid in every $\mathfrak{F} \in \mathfrak{C}$ (notation: $X \vdash_{\mathfrak{C}} A$).

A (positive) substructural logic L is *characterised* by a class of frames $\mathfrak{C}$ iff the following holds: $X \vdash_{\mathfrak{C}} A$ iff $X \vdash A$ is provable in L. (In general, 'logics' are seen as sets of consecutions and 'provable in L' is construed accordingly as 'being a member of L'. In what follows, we take the logics to be sets of consecutions derivable in specific natural deduction systems in the style of [10, Ch. 2]. The systems are given by a set of axioms ($A \vdash A$ for every formula A), introduction and elimination rules for every connective and structural rules).

Definition 2 A *pair frame* is a triple

$$\mathbf{F} = (W, R_0, \{R^i_j\})_{i,j \in \{1,2\}}$$

where W is a non-empty set and R_0, R^i_j are binary relations on W.

A *pair model* built on a frame $\mathbf{F}$ is a couple

$$\mathbf{M} = (\mathbf{F}, \{V_i\})_{i \in \{1,2\}}$$

such that $\mathbf{F}$ is a pair frame and V_i are valuations, i.e. functions from Φ to subsets of W.

A pair model is a set of points together with five binary relations and two valuations. Hence, a pair model $\mathbf{M}$ might be seen as a pair of multimodal Kripke models $\langle \mathbf{M}_1, \mathbf{M}_2 \rangle$. For example:

- $\mathbf{M}_1 = (W, R_0, R^1_1, R^1_2, V_1)$
- $\mathbf{M}_2 = (W, R^2_1, R^2_2, V_2)$

A pair frame is simply a multimodal Kripke frame. (But, on the other hand, it may be seen as a pair of Kripke frames as well.)

Definition 3 The valuations V_1, V_2 give rise to two truth relations $\models_1$ and $\models_2$ ($i \in \{1, 2\}$):

- $(\mathbf{M}, w) \models_i p$ iff $w \in V_i(p)$
- $(\mathbf{M}, w) \models_i A \wedge B$ iff $(\mathbf{M}, w) \models_i A$ and $(\mathbf{M}, w) \models_i B$
- $(\mathbf{M}, w) \models_i A \vee B$ iff $(\mathbf{M}, w) \models_i A$ or $(\mathbf{M}, w) \models_i B$
- $(\mathbf{M}, w) \models_i A \rightarrow B$ iff $R^i_1 wv$, $(\mathbf{M}, v) \models_2 A$ and $R_0 vu$ imply $(\mathbf{M}, u) \models_1 B$, for all $v, u \in W$.
- $(\mathbf{M}, w) \models_i A \circ B$ iff there are $v, u \in W$ such that $R^i_1 wv$, $R_0 uv$, $(\mathbf{M}, u) \models_1 A$, and $(\mathbf{M}, u) \models_2 B$.
- $(\mathbf{M}, w) \models_i B \leftarrow A$ iff $R^i_2 wv$, $(\mathbf{M}, v) \models_1 A$ and $R_0 vu$ imply $(\mathbf{M}, u) \models_1 B$, for all $v, u \in W$.

These conditions are extended to structures similarly as it is done in the context of substructural models (see [10]). Hence, ';' mimics '$\circ$' while ',' mimics '$\wedge$'. A consecution $X \vdash A$ is *valid in* $\mathbf{M}$ iff $(\mathbf{M}, w) \models_1 X$ implies $(\mathbf{M}, w) \models_1 A$, for all $w \in W$ (notation: $X \vdash_{\mathbf{M}} A$).

A consecution $X \vdash A$ is *valid in a frame* $\mathbf{F}$ iff it is valid in every $\mathbf{M}$ built on $\mathbf{F}$ (notation: $X \vdash_{\mathbf{F}} A$). If $\mathbf{C}$ is a class of pair frames, then $X \vdash A$ is *valid in* $\mathbf{C}$ iff it is valid in every $\mathbf{F} \in \mathbf{C}$ (notation: $X \vdash_{\mathbf{C}} A$).

A (positive) substructural logic L is *characterised* by a class of pair frames $\mathbf{C}$ iff the following holds: $X \vdash_{\mathbf{C}} A$ iff $X \vdash A$ is provable in L.

Definition 4 A pair frame $\mathbf{F}$ is a *pair frame representation* (a *p.f.r.*) of a substructural frame $\mathfrak{F}$ iff there is a bijection $\sigma : W \to P^2$ such that (w_i denotes the i-th member of $\sigma(w)$):

- R_0wv iff $Rw_1w_2v_1$ and
- R^i_jwv iff $w_i = v_j$

A pair model $\mathbf{M} = (\mathbf{F}, \{V_i\})_{i,j\in\{1,2\}}$ is a *pair model representation* (a *p.m.r.*) of a substructural model $\mathfrak{M} = (\mathfrak{F}, \Vdash)$ iff:

- $\mathbf{F}$ is a *p.f.r.* of $\mathfrak{F}$ and
- $w \in V_i(p)$ iff $w_i \Vdash p$

If $\mathfrak{S}$ is a substructural model, frame or a class of frames, then the respective class of pair representations of $\mathfrak{S}$ is denoted $\mathsf{Rep}(\mathfrak{S})$.

Note that $\mathsf{Rep}(\mathfrak{F})$, $\mathsf{Rep}(\mathfrak{M})$ are non-empty, for every $\mathfrak{F}$, $\mathfrak{M}$. In addition, observe that if $\mathbf{M}$ is built upon $\mathbf{F}$ and $\mathbf{F} \in \mathsf{Rep}(\mathfrak{F})$, then $\mathbf{M} \in \mathsf{Rep}(\mathfrak{M})$ for some $\mathfrak{M}$ built on $\mathfrak{F}$.

A *p.m.r.* of $\mathfrak{M}$ represents the information contained in $\mathfrak{M}$ by means of a pair of multimodal Kripke models. The first step is to rephrase the substructural model in terms of pair points: the 'new' points are pairs $\langle xy\rangle$ of the 'old' points. From this point of view, the ternary R may be replaced by a binary R' such that $R'\langle xy\rangle\langle zz'\rangle$ iff $Rxyz$. Now the points $w \in \mathbf{M}$ can be seen as representations of the pairs $\langle xy\rangle$, if there is a bijection that preserves their properties and their 'position' among other pairs. Of course, the position and properties are given by (i) the relation R', (ii) the inner structure of the pairs and (iii) the valuation. Now R_0 'models' the binary R' and, therefore, the ternary R. The relations R^i_j are there to 'keep track' of the inner structure of the modelled pairs. The notation itself suggests that R^i_jwv means that the i-th member of the pair represented by w (i.e. of $\sigma(w)$) is identical with the j-th member of the pair represented by v (i.e. of $\sigma(v)$). Last but not least, the valuations reflect the 'binary nature' of pair points (the principle is explained in Sect. 2.2).

It is now obvious that the seemingly awkward truth conditions for formulas containing '$\to$', '$\circ$', '$\leftarrow$' are 'mere' reformulations of the corresponding truth conditions in substructural models. As we shall see in Sect. 3.3, this yields a general result about the characterisation of (positive) substructural logics by means of pair frames.

3.2 Implicit Pair Point Structures

But first, let us pause and consider the following question. Pair point (frame) representations of substructural models (frames) are, quite understandably, defined relatively to substructural models (frames). However, one might ask if there are 'independent' means of identifying pair representations. This section answers affirmatively.

Definition 5 A pair frame **F** is an *implicit pair point frame* (*i.p.p.* frame) iff there is a bijection $\beta : W_{\mathbf{F}} \to S^2$ for some non-empty S such that $R^i_j wv$ iff $w_i = v_j$. (w_i (v_j) is the ith (jth) member of $\beta(w)$ ($\beta(v)$), for $i, j \in \{1, 2\}$.)

A pair model $\mathbf{M} = (\mathbf{F}, \{V_i\})_{i \in \{1,2\}}$ is an *implicit pair point model* (*i.p.p.* model) iff (i) **F** is an *i.p.p.* frame and (ii) $R^i_j wv$ implies ($w \in V_i(p)$ iff $v \in V_j(p)$) for all $p \in \Phi$.

Proposition 1 *A pair frame is a p.f.r. of some $\mathfrak{F}$ iff it is an i.p.p. frame. A pair model is a p.m.r. of some $\mathfrak{M}$ iff it is an i.p.p. model.*

Proof First, let us consider frames. The implication from left to right is trivial. The converse implication follows from the following construction. Let **F** be an *i.p.p.* frame and let $\mathfrak{F} = (S, \sqsubseteq, R)$ such that $\sqsubseteq$ is a partial order on S and $Rxyz$ iff $x = w_1$, $y = w_2$, $z = v_1$ and $R_0 wv$. Obviously, **F** is a *p.f.r.* of $\mathfrak{F}$.

Next, let us consider models. Again, the left-to-right implication is trivial. The converse implication follows from considering a model $\mathfrak{M} = (S, \sqsubseteq, R, \Vdash)$ such that R is defined as above, $x \Vdash p$ iff $x = w_i$ and $w \in V_i(p)$. Moreover, let $\sqsubseteq$ be a partial order on S such that $x \Vdash p$ and $x \sqsubseteq y$ imply $y \Vdash p$. □

3.3 Pair Frames and Substructural Logics

Lemma 1 *Let* **M** *be a p.m.r. of $\mathfrak{M}$. Then*

$$(\mathbf{M}, w) \models_i A \textit{ iff } (\mathfrak{M}, w_i) \Vdash A$$

for every bijection σ with the properties specified in Definition 4. The same holds for structures.

Proof The basic case $A = p$ holds by Definition 4. The cases $A = B \wedge C, A = B \vee C$ are trivial.

Next, assume that $w \not\models_i B \to C$. This means that there are v, u such that $R^i_1 wv$, $R_0 vu$, $v \models_2 B$, but $u \not\models_1 C$. By the induction hypothesis and by Definition 4, this amounts to $w_i = v_1$, $Rv_1v_2u_1$, $v_2 \Vdash B$, but $u_1 \nVdash C$. However, this holds iff $w_i \nVdash B \to C$. This completes the proof for case $A = B \to C$. The cases $A = B \circ C$ and $A = C \leftarrow B$ are proved similarly. The cases for structures are virtually identical to cases for $\wedge$ and $\circ$. □

Lemma 2

(a) $X \vdash_{\mathfrak{M}} A$ *iff* $X \vdash_{\mathsf{Rep}(\mathfrak{M})} A$
(b) $X \vdash_{\mathfrak{F}} A$ *iff* $X \vdash_{\mathsf{Rep}(\mathfrak{F})} A$
(c) $X \vdash_{\mathfrak{C}} A$ *iff* $X \vdash_{\mathsf{Rep}(\mathfrak{C})} A$

Proof (a) Assume that $X \nvdash_{\mathfrak{M}} A$. There is a point $x \in \mathfrak{M}$ such that $(\mathfrak{M}, x) \Vdash X$, but $(\mathfrak{M}, x) \nVdash A$. Let **M** be any *p.m.r.* of $\mathfrak{M}$ and let $\sigma(w) = \langle xy \rangle$ for some $y \in \mathfrak{M}$. By Lemma 1, $(\mathbf{M}, w) \models_1 X$, but $(\mathbf{M}, w) \not\models_1 A$. Consequently, $X \nvdash_{\mathsf{Rep}(\mathfrak{M})} A$.

Now assume that $X \not\models_{\mathsf{Rep}(\mathfrak{M})} A$, i.e. there is a *p.m.r.* $\mathbf{M}$ of $\mathfrak{M}$ such that $(\mathbf{M}, w) \models_1 X$, but $(\mathbf{M}, w) \not\models_1 A$ for some $w \in \mathbf{M}$. By Lemma 1, $(\mathfrak{M}, w_1) \Vdash X$, but $(\mathfrak{M}, w_1) \nVdash A$. Hence, $X \not\models_{\mathfrak{M}} A$.

(b) Assume that $X \not\models_{\mathfrak{F}} A$. There is a model $\mathfrak{M} = (\mathfrak{F}, \Vdash)$ and a point x such that $(\mathfrak{M}, x) \Vdash X$, but $(\mathfrak{M}, x) \nVdash A$. By Lemma 1, every $\mathbf{M} \in \mathsf{Rep}(\mathfrak{M})$ invalidates $X \vdash A$ as well. By Definition 4, if such $\mathbf{M}$ is built on a frame $\mathbf{F}$, then $\mathbf{F} \in \mathsf{Rep}(\mathfrak{F})$. Hence $X \not\models_{\mathsf{Rep}(\mathfrak{F})} A$.

Assume $X \not\models_{\mathsf{Rep}(\mathfrak{F})} A$. There is a model $\mathbf{M}$ built on a frame $\mathbf{F}$ and a point w such that $\mathbf{F} \in \mathsf{Rep}(\mathfrak{F})$, $(\mathbf{M}, w) \models_1 X$, but $(\mathbf{M}, w) \not\models_1 A$. By the remark following Definition 4, there is a substrucutral $\mathfrak{M} = (\mathfrak{F}, \Vdash)$, such that $\mathbf{M} \in \mathsf{Rep}(\mathfrak{M})$. By Lemma 1, $X \not\models_{\mathfrak{M}} A$. Consequently $X \not\models_{\mathfrak{F}} A$.

(c) is an immediate consequence of (b). □

Theorem 1 (General Pair-Frame Theorem) *If a (positive) substructural logic L is characterised by a class of substructural frames $\mathfrak{C}_L$, then it is characterised by the class of pair frames* $\mathsf{Rep}(\mathfrak{C}_L)$.

Proof Follows from Lemma 2 (c). □

Hence, many well-known positive substructural logics are characterised by multi-modal Kripke frames.

In conjunction with Proposition 1, Theorem 1 yields a specific soundness and completeness result for some very weak substructural logics.

Theorem 2 *A consecution $X \vdash A$ is valid in every $\mathfrak{F}$ iff it is valid in every i.p.p. model.*

3.4 Inter-Model Connectives

Theorem 1 is not supposed to suggest that '$\rightarrow$', '$\circ$', '$\leftarrow$' are 'on a par' with the usual modal-like binary operators (e.g., see [4]). The truth conditions of formulas containing '$\rightarrow$', '$\circ$', '$\leftarrow$' are obviously not ordinary modal clauses. Note again that pair models $\mathbf{M}$ can be seen as pairs of Kripke models $\langle \mathbf{M}_1, \mathbf{M}_2 \rangle$. Moreover, observe that, for example, the condition for $\models_1 A \rightarrow B$ refers also to $\models_2$ (similarly for '$\circ$', '$\leftarrow$'). This shows that, in general, formulas with '$\rightarrow$', '$\circ$', '$\leftarrow$' 'operate' *between* the models $\mathbf{M}_1$ and $\mathbf{M}_2$ in the pair model (or 'pair of models') $\mathbf{M}$.[3]

This yields a 'hierarchy of operators': boolean operators operate 'within' points in models; modal operators operate 'between' points, but always 'within' models; substructural operators operate between points and between models.

[3] A familiar example of similar 'inter-model' truth conditions are the conditions for *public announcement* formulas $[A]B$ in public announcement logic. See [11].

4 An Independent Completeness Result

Section 3 contains two characterisation results. Theorem 1 is quite general, but it refers to structures that are 'constructed from' substructural frames. Theorem 2 establishes completeness of some very weak substructural logics with respect to *i.p.p.* models and their definition does not directly refer to substructural models. We demonstrate in this section that *i.p.p.* models are perhaps too restrictive: we obtain a similar completeness result with respect to a broader class of pair models. Interestingly, not every member of this broader class is a pair model representation of some substructural model.

Definition 6 A pair model **M** is *substructural* iff ($i, j, k \in \{1, 2\}$):

(a) $R^i_j wv$ implies ($R^j_k vu$ iff $R^i_k wu$) and
(b) $R^i_i ww$ and
(c) $\forall w \exists v R^i_j wv$ and
(d) $R^i_j wv$ implies ($w \in V_i(p)$ iff $v \in V_j(p)$)

for all $w, u, v \in \mathbf{M}$ and $p \in \Phi$.

Lemma 3 *Every i.p.p. model is a substructural pair model. However, the converse does not hold.*

Proof By Def. 5, if **M** is an *i.p.p.* model, then there is a bijection $\beta : W_{\mathbf{M}} \to S^2$ for some non-empty S such that $R^i_j wv$ iff $w_i = v_j$. Hence, if $R^i_j wv$ and $R^j_k vu$, then $w_i = v_j = u_k$ and, consequently $R^i_k wu$. The rest of item (a) is proved similarly. Item (b) is a trivial consequence of the reflexivity of identity, and items (c), (d) are immediate consequences of Def. 5.

To prove the remaining claim of the Lemma, assume that we have a substructural pair model **M** such that $W_{\mathbf{M}} = \{w, v\}$. It is plain that, for arbitrary non-empty S, there is no bijection from $W_{\mathbf{M}}$ to S^2, since $|S^2| \neq 2$ for every S. □

Proposition 2 *Every substructural logic L characterised by the class of all frames is complete with respect to the class of substructural pair models.*

Proof Follows from Proposition 1, Theorem 2 and Lemma 3. □

Definition 7 An inference rule

$$X_1 \vdash A_1, \ldots, X_n \vdash A_n \implies X \vdash A$$

is *admissible* in a class of pair models iff every model in the class that validates the consecutions $X_1 \vdash A_1, \ldots, X_n \vdash A_n$ validates the consecution $X \vdash A$ as well.

To prove soundness, we have to demonstrate that every introduction and elimination rule for the connectives of $\mathscr{L}^+$ is admissible in the class of substructural pair models.

Lemma 4 *Let* **M** *be a substructural pair model. If* $R^i_j wv$*, then*

$$(\mathbf{M}, w) \models_i A \quad \textit{iff} \quad (\mathbf{M}, v) \models_j A$$

The same holds for structures.

Proof The basic case $A = p$ holds by Definition 6. The cases $A = B \wedge C$ and $A = B \vee C$ are trivial.

Next, assume that $w \not\models_i B \rightarrow C$. This amounts to: $\exists uu'$ such that $R^i_1 wu$, $R_0 uu'$, $u \models_2 B$ and $u' \not\models_1 C$. By Definition 6, if $R^i_j wv$, then $R^j_1 vu$ is equivalent to $R^i_1 wu$. Hence, $w \not\models_i B \rightarrow C$ iff $v \not\models_j B \rightarrow C$. The cases $A = B \circ C$ and $A = C \leftarrow B$ are proved similarly (use case a) of Definition 6). □

Lemma 5 *Let* **M** *be a substructural pair model. If* $X \vdash_{\mathbf{M}} A$ *and* $w \models_2 X$ *for some* $w \in \mathbf{M}$*, then* $w \models_2 A$*.*

Proof Assume that $X \vdash_{\mathbf{M}} A$ and $w \models_2 X$ for some $w \in \mathbf{M}$. By Definition 6, $\exists v$ such that $R^2_1 wv$. By Lemma 4, the assumption implies that $v \models_1 X$. Consequently, $v \models_1 A$. By Lemma 4 again, $w \models_2 A$. □

In the following Lemma, $Y(X)$ means that X occurs at least n times in Y as a substructure ($n \leq 0$). $Y(X/A)$ is the result of replacing every occurrence of X in Y by an occurrence of A (similarly for $Y(A/X)$, see [10]).

Lemma 6 *If* $X \vdash_{\mathbf{M}} A$ *and* $(\mathbf{M}, w) \models_1 Y(X)$*, then* $(\mathbf{M}, w) \models_1 Y(X/A)$*.*

Proof The proof is by induction on the complexity of Y. If Y does not contain X, then the claim holds vacuously. If $Y = X$, then $w \models_1 Y(X)$ implies $w \models_1 Y(X/A)$ according to the assumption $X \vdash_{\mathbf{M}} A$. Now assume that $w \models_1 Z(X); Z'(X)$. Hence, we have points v, u such that $R^1_1 wv$, $R_0 uv$, $u \models_1 Z(X)$ and $u \models_2 Z'(X)$. By Lemma 5 and the induction hypothesis, $u \models_2 Z'(A)$. By another application of the induction hypothesis, $u \models_1 Z(A)$. Hence, $w \models_1 Z(A); Z'(A)$. The claim for $Z(X), Z'(X)$ follows from the induction hypothesis. □

Lemma 7 *Every introduction and elimination rule for the connectives of* $\mathscr{L}^+$ *(see [10, Ch. 2]) is admissible in the class of substructural pair models.*

Proof We shall prove the claim for $\rightarrow$-elimination

$$X \vdash A \rightarrow B, Y \vdash A \Longrightarrow X; Y \vdash B$$

and $\circ$-elimination

$$X \vdash A \circ B, Y(A; B) \vdash C \Longrightarrow Y(A; B/X) \vdash C$$

The other parts of the proof are analogous.

Assume that $X \vdash_{\mathbf{M}} A \to B$ and $Y \vdash_{\mathbf{M}} A$ for some substructural pair model $\mathbf{M}$. Moreover, assume that $w \models_1 X; Y$ for some $w \in \mathbf{M}$. The latter assumption amounts to: $\exists vu$ such that $R^1_1 wv$, $R_0 uv$, $u \models_1 X$ and $u \models_2 Y$. By the former assumption, $u \models_1 A \to B$. By Lemma 5, $u \models_2 A$. By Definition 6, $R^1_1 uu$. Consequently, $v \models_1 B$. By Lemma 4, $w \models_1 B$.

Next, assume that $X \vdash_{\mathbf{M}} A \circ B$, $Y(A; B) \vdash_{\mathbf{M}} C$ and $w \models_1 Y(A; B/X)$. By Lemma 6, the assumptions imply $w \models_1 Y(A \circ B)$. However, this amounts to $w \models_1 Y(A \circ B/A; B)$. By the third assumption, $w \models_1 C$. □

Theorem 3 (Direct Characterisation Theorem) *Every substructural logic L characterised by the class of all substructural frames is sound and complete with respect to the class of all substructural pair models.*

Proof Follows from the obvious fact that $A \vdash_{\mathbf{M}} A$ for every pair model $\mathbf{M}$, Proposition 2 and Lemma 7. □

5 An Interpretation of Pair Models

The pair model semantics seems to be rather awkward and in need of clarification. This section offers a 'philosophical story' behind the semantics. The story uses epistemic notions, mainly related to communication and inference. The story results in an epistemic reading of the substructural connectives.

The presence of two valuations in pair models shall be our starting point. The valuations can be given an epistemic reading as follows. (Subsequently, $i, j \in \{1, 2\}$.) As usual, we can see points $w \in W$ as situations (worlds, time instants etc.). The valuations V_1, V_2 are seen as corresponding to two agents *agent*$_1$ and *agent*$_2$ (a_1, a_2 for short). More specifically, if w^*_i is the set $\{p \mid w \in V_i(p)\}$, then w^*_i is seen as the set of *atomic information* available to a_i at w. Consequently, the valuations V_i describe the sets of atomic information available to the agents a_i at situations $w \in W$.

The relations $\models_i$ are seen as specifying *confirmation* conditions for (complex) bodies of information. More specifically, $w \models_i A$ may be read as 'the information available to a_i at w confirms A'. (Notation: $\mathsf{Con}_i(w) = \{A \mid w \models_i A\}$.)

From a more general perspective, the various $w \in W$ correspond to different 'set-ups' in terms of the atomic information available to agents a_i. Now assume that a_1, a_2 are not proper names (rigid designators), but 'situation-relative' tags. Hence, every situation w is assigned a pair of agents $a_i(w)$. Consequently, every w may be seen as an *interaction set-up*: the order of the agents matters. In other words, it is expected that the role of $a_i(w)$ in the interaction set-up w varies with different values of i. This idea will be explored shortly.

The above reading of situations as interaction set-ups consisting of two agents yields a natural interpretation of the relations R^i_j. These serve the purpose of 'cross-situational identification' of agents. More specifically, $R^i_j wv$ may be read as 'the agent $a_i(w)$ is identical with $a_j(v)$. This provides an interpretation of substructural pair models (see Definition 6) and, consequently, pair model simulations. In these pair models, $R^i_j wv$ implies $w^*_i = v^*_j$. In other words, the set of atomic information available to agents does not change across situations. And, as Lemma 4 witnesses, nor do the sets of confirmed information $\mathsf{Con}_i(w)$.

However, change and processing of available information plays a vital role in our interpretation. The relation R_0 may be seen as describing possible outcomes of communication and inference. In other words, $R_0 wv$ may be read as follows: Suppose that the agent $a_2(w)$ provides the information $\mathsf{Con}_2(w)$ to the agent $a_1(w)$. The latter agent processes the information $\mathsf{Con}_1(w) \cup \mathsf{Con}_2(w)$ and (by means of inference, among others) arrives at the body of information $\mathsf{Con}_1(v)$. In other words, $R_0 wv$ may mean that the set-up v is a possible result of communication and inference 'within' the set-up w. We shall say in this case that w is a *source* of v.

The specific truth conditions are now easy to spell out. A proposition letter p is confirmed by $a_i(w)$ iff p is among the atomic information available to $a_i(w)$ at w. A conjunction $A \wedge B$ is confirmed iff both conjuncts are confirmed. A disjunction is confirmed iff at least one of the disjuncts is confirmed (i.e. the agents behave 'intuitionistically').

Let us call the agent $a_2(w)$ the *sender* (of w) and $a_1(w)$ the *reasoner* (of w). A conditional $A \rightarrow B$ is confirmed by $a_i(w)$ in w (let us name the agent a) iff the following holds: if a was the reasoner in an interaction set-up v where the sender confirms A and if a engaged in inference from the assumptions $\mathsf{Con}_1(v) \cup \mathsf{Con}_2(v)$, then a would confirm B. In other words, we arrive at the familiar epistemic interpretation of the substructural conditional (at least if we assume that we are working with substructural pair models): $A \rightarrow B$ is confirmed with respect to a body of information iff extending the body of information by A yields B.[4] The interpretation of the converse conditional $B \leftarrow A$ is similar and may be easily derived from the interpretation of $A \rightarrow B$.

Fusion $A \circ B$ is confirmed by $a_i(w)$ in w (a again) iff there is a possible interaction set-up v, where i) a is the reasoner and ii) v is the result of communication and inference within a set-up u (i.e. $R_0 uv$ holds) such that $A \in \mathsf{Con}_1(u)$ and $B \in \mathsf{Con}_2(u)$. Observe that, if we are working with substructural pair models, this yields $A \circ B \in \mathsf{Con}_1(v)$ as well as $A \circ B \in \mathsf{Con}_i(w)$. In other words, $A \circ B$ is confirmed with respect to a body of information iff there is a source of the information that confirms A as well as B. (Of course, order does matter: commutativity does not hold in general.)

6 Conclusion

The paper developed the idea of pair points [3] into a fully fledged binary semantics for (positive) substructural logics in terms of 'pair frame (model) representations' of substructural frames (models). A general characterisation result has been provided: if a (positive) substructural logic is characterised by a class of frames $\mathfrak{C}$, then it is

[4] Needless to point out, the interpretation is rather close to the Ramsey Test.

characterised by the class $\mathsf{Rep}(\mathfrak{C})$ of pair frame representations of frames $\mathfrak{F} \in \mathfrak{C}$. As an interesting aside, an 'independent characterisation' of pair representations as well as an 'independent characterisation result' have been established: every substructural logic characterised by the class of all frames is characterised by the class of substructural pair models. Hence, we have directly identified the class of pair models (and frames) that characterises some very weak substructural logics. It has been pointed out that these results indicate that substructural connectives fit in a hierarchy, along with boolean and modal operators. The latter two operate inside points and ordinary modal models respectively, while substructural connectives operate between modal models. Hence, they have an inter-model nature.

However, many issues remain open. First, we should be able to incorporate negation into our semantics. This should not be very hard to do. In fact, it is expected that such extensions will be dealt with in an extended version of this paper. Of course, this advance would yield general characterisation results for a more comprehensive class of substructural logics (not only the positive ones).

Second, as we have already noted, a characterisation result in the style of Theorem 1 is rather indirect. For every L, the class of pair frames $\mathsf{Rep}(\mathfrak{C}_L)$ is 'constructed from' the frames in $\mathfrak{C}_L$. Hence, 'direct descriptions' of classes of pair frames corresponding to individual substructural logics are much desired.

Third, the 'philosophical story' behind the pair semantics may be seen as somewhat sketchy. Hence, a deeper and more comprehensive version should be provided. However, these investigations are left for another occasion.

Acknowledgments This work was carried out at the Department of Logic and Methodology of Sciences, Comenius University, as a part of the research project 'Semantic models, their explanatory power and applications', supported by the grant VEGA 1/0046/11. I wish to express my gratitude to the audience at Trends in Logic XI for helpful discussion and to two anonymous referees for enabling me to improve the paper by providing a number of constructive suggestions. Thanks are also due to Jc Beall, Mike Dunn, Ed Mares and Graham Priest: their encouragement is much appreciated.

References

1. Barwise, J., & Perry, J. (1983). *Situations and attitudes*. Cambridge: MIT Press.
2. Beall J. C. (2009) *Spandrels of truth*. Oxford: Oxford University Press.
3. Beall, J. C., Brady, R., Dunn, J. M., Hazen, A., Mares, E., Meyer, R., et al. (2012). On the ternary relation and conditionality. *Journal of Philosophical Logic*, *41*(3), 595–612.
4. Blackburn, P., Rijke, M., & Venema, Y. (2001). *Modal logic*. Cambridge: Cambridge University Press.
5. Dunn, J. M. (1976). A Kripke-style semantics for R-mingle using a binary accessibility relation. *Studia Logica*, *35*(2), 163–172.
6. Dunn, J. M. (1987). Incompleteness of the bibinary semantics for R. *The Bulletin of the Section of Logic*, *16*(3), 107–109.
7. Kurtonina, N. (1998). Categorial inference and modal logic. *Journal of Logic, Language and Information*, *7*, 399–411.
8. Mares, E. (2004). *Relevant logic: A philosophical interpretation*. Cambridge: Cambridge University Press.

9. Meyer, R. K., & Routley, R. (1973). Classical relevant logics II. *Studia Logica*, *33*(2), 183–194.
10. Restall, G. (2000). *An introduction to substructural logics*. London and New York: Routledge.
11. van Ditmarsch, H., van der Hoek, W., & Kooi, B. (2008). *Dynamic epistemic logic*. Dordrecht: Springer.
12. Vasyukov, V. L. (1986). The bibinary semantics for R and $Ł_{\aleph_0}$. *The Bulletin of the Section of Logic*, *15*(3), 109–114.
13. Vasyukov, V. L. (1994). From ternary to tetrary? *The Bulletin of the Section of Logic*, *23*(4), 163–167.

MIX
Papier aus verantwortungsvollen Quellen
Paper from responsible sources
FSC® C105338

If you have any concerns about our products,
you can contact us on
ProductSafety@springernature.com

In case Publisher is established outside the EU,
the EU authorized representative is:
Springer Nature Customer Service Center GmbH
Europaplatz 3, 69115 Heidelberg, Germany

Printed by Libri Plureos GmbH
in Hamburg, Germany